ADVANCES IN ENZYMOLOGY

AND RELATED AREAS OF
MOLECULAR BIOLOGY

Volume 70

LIST OF CONTRIBUTORS

SHIN-ICHI AOTA, Laboratory of Developmental Biology, National Institute of Dental Research, National Institutes of Health, Bethesda, MD 20892.

RAYMOND L. BLAKLEY, Department of Molecular Pharmacology, St. Jude Children's Research Hospital, Memphis, TN 38101.

A. M. CHAKRABARTY, Department of Microbiology and Immunology (M/C 790), University of Illinois College of Medicine, Chicago, IL 60612.

SEYMOUR KAUFMAN, Laboratory of Neurochemistry, National Institute of Mental Health, National Institutes of Health, Bethesda, MD 20892.

DAVID SCHLICTMAN, Department of Microbiology and Immunology (M/C 790), University of Illinois College of Medicine, Chicago, IL 60612.

SANDEEP SHANKAR, Department of Microbiology and Immunology (M/C 790), University of Illinois College of Medicine, Chicago, IL 60612.

KENNETH M. YAMADA, Laboratory of Developmental Biology, National Institute of Dental Research, National Institutes of Health, Bethesda, Maryland 20892.

RICK W. YE, Department of Microbiology and Immunology (M/C 790), University of Illinois College of Medicine, Chicago, IL 60612.

ADVANCES IN ENZYMOLOGY
AND RELATED AREAS OF MOLECULAR BIOLOGY

Founded by F. F. NORD

Edited by ALTON MEISTER

CORNELL UNIVERSITY MEDICAL COLLEGE
NEW YORK, NEW YORK

VOLUME 70

WILEY
1995

AN INTERSCIENCE® PUBLICATION

JOHN WILEY & SONS, INC.
New York • Chichester • Brisbane • Toronto • Singapore

This text is printed on acid-free paper.

This publication is designed to provide accurate and authoritative information in regard to the subject matter covered. It is sold with the understanding that the publisher is not engaged in rendering legal, accounting, or other professional services. If legal advice or other expert assistance is required, the services of a competent professional person should be sought.

Library of Congress Catalog Card Number: 41-9213
ISBN 0-471-04097-5

Printed in United States of America

10 9 8 7 6 5 4 3 2 1

CONTENTS

v

ADVANCES IN ENZYMOLOGY

AND RELATED AREAS OF
MOLECULAR BIOLOGY

Volume 70

FIBRONECTIN AND CELL ADHESION: SPECIFICITY OF INTEGRIN–LIGAND INTERACTION

By SHIN-ICHI AOTA
and KENNETH M. YAMADA, *Laboratory of Developmental Biology, National Institute of Dental Research, National Institutes of Health, Bethesda, Maryland 20892*

CONTENTS

I. Introduction

Fibronectin is a large, extracellular glycoprotein with many biological activities. One activity of great interest involves its adhesiveness for cells via integrin cell–surface receptors. Integrins represent a large group of heterodimeric receptors that are of central importance in cell–surface interactions between many cell types and a variety of extracellular matrix protein ligands including fibronectin. These interactions between integrins and their ligands play important roles in a vast array of biological processes, including cell adhesion and migration, wound healing, embryogenesis, and signal transduction.

Advances in Enzymology and Related Areas of Molecular Biology, Volume 70, Edited by Alton Meister.
ISBN 0-471-04097-5 © 1995 John Wiley & Sons, Inc.

The cell-binding domain of fibronectin has been the most extensively studied region among the many sites in extracellular matrix proteins known to interact with integrin receptors and to mediate cell adhesion and migration. This region consists of repeating units approximately 90 amino acid residues in length, termed fibronectin type III repeats (1). A key attachment site for integrins, the Arg-Gly-Asp (RGD) cell-adhesion motif, was identified in this domain in the 10th type III repeating unit of fibronectin. Synthetic peptides containing this RGD sequence can mimic or competitively inhibit the cell-adhesive activity of this cell-binding domain of fibronectin. Subsequent studies of other extracellular matrix proteins has led to the identification of many such short binding sites, including the RGD sequence in vitronectin and the Leu-Asp-Val (LDV) sequence in the IIICS region of fibronectin.

Nearly one-half of the more than 20 members of the integrin family are known to interact with RGD sequences in their target proteins, including the integrins $\alpha 5 \beta 1$, $\alpha IIb \beta 3$, and $\alpha v \beta 3$. However, these RGD-dependent integrins bind to their RGD-containing ligands such as fibronectin, vitronectin, and fibrinogen with distinctively different affinities and specificity. For example, the integrin $\alpha 5 \beta 1$ is the primary receptor for the cell-binding domain of fibronectin in many cell types, yet it binds poorly to vitronectin. Moreover, RGD-containing peptides usually show decreased and less-specific activity compared to the cell-binding domain of fibronectin. In fact, many proteins that contain the RGD motif such as immunoglobulins and β-galactosidase are not adhesive proteins.

These observations led to studies on the effects of conformation of RGD peptides, as well as to a search for additional sequence information besides the RGD motif that is important for full cell-adhesive activity. In the first series of studies, the most notable finding is that conformational constraint of RGD-containing peptides can result in increased affinity for certain integrins, especially for members of the β_3 integrin subfamily. In the second type of approach, recent studies indicated that additional polypeptide information besides the RGD sequence is required for full cell-adhesive activity. In addition, three-dimensional (3-D) structures have been determined for an increasing number of cell-adhesive protein domains. These data combine to provide important clues about the mechanisms of integrin–ligand interaction.

Recently, there have been many reviews on fibronectin and integrin-mediated cell adhesion. They include general reviews on fibronectin structure and functions (2–4), on cell-adhesive sequences in extracellular matrix proteins (5, 6), and on integrins and their ligands (7–9). This chapter concentrates on the recent advances in the study of integrin ligands and on how they interact with integrins, rather than on integrin receptors per se. Studies of fibronectin provided important insights into the mechanisms of these interactions.

II. Integrin Cell–Surface Receptors

Integrins are heterodimeric membrane glycoproteins consisting of two subunits, α and β. Each subunit has a transmembrane domain, a large extracellular domain, and a small (except β4) cytoplasmic domain. There are at least 14α subunits and 8β subunits in vertebrate cells, in addition to several other α and β subunits identified in other species, such as in *Drosophila melanogaster* and *Caenorhabditis elegans*. Figure 1 shows the known combinations of α and β subunits and their ligands. One notable feature of integrin–ligand interactions is frequent promiscuity; many integrins can bind to multiple ligands, which are often evolutionarily and structurally unrelated to each other. This complexity arises at least partly from the fact that integrins can recognize short amino acid sequences present in many proteins.

III. Short Integrin-Recognition Sequences

Past studies using synthetic peptides have identified many short amino acid sequences with which integrins interact (Table I). At least eight integrins recognize the RGD sequence, but with various affinities. Other integrin-binding sequences include KQAGDV and LDV, which have been identified as the binding site for integrin αIIbβ3 in the fibrinogen γ chain, and the α4β1 binding site in an alternatively spliced region (IIICS) of fibronectin, respectively [reviewed in (5)].

One notable recent trend has been that the list of interacting ligand sequences for each integrin member has started to expand. For example, the α5β1 integrin, for which the RGD sequence had originally been thought to be the sole binding sequence, now has several interacting sequences according to competitive inhibition studies (Table I). In addition to these peptide interactions, a protein termed invasin, which is an outer-membrane protein of *Yersinia pseudotuberculosis*,

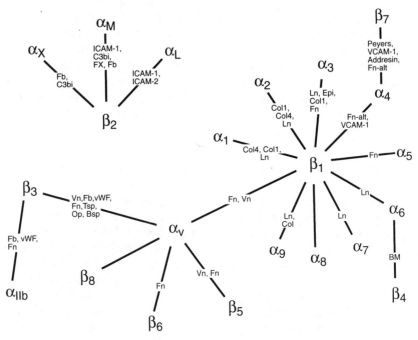

Figure 1.   Integrin subunits and ligands. The known pairings of integrin α and β subunits to form the final integrin heterodimers are depicted by the lines, on which are listed the specific ligand(s) bound by that heterodimer. The abbreviations for ligands are Fn = fibronectin central cell-binding region; Fn-alt = fibronectin alternatively spliced region; Ln = laminin; Epi = epiligrin (kalinin); Fb = fibrinogen; Coll = collagen type I; Col4 = collagen type IV; vWF = von Willebrand factor; Op = osteopontin; Tsp = thrombospondin; C3bi = complement component C3bi; Peyers = Peyers patch; FX = factor X; Bsp = bone sialoprotein 1.

and a protein termed disagregin from the tick *Ornithodoros moubata* have also been shown to bind β1 integrins and αIIbβ3, respectively, even though their binding domains do not contain an RGD sequence (34, 35). Integrins may therefore sometimes have quite relaxed specificities for interactions with various sequences with differing affinities. General screening methods for studying interactions and binding, such as the use of phage display libraries (18, 36) and characterization of new cell-adhesive proteins like invasin, are likely to identify even more new motifs for integrin interaction sites.

TABLE I
Integrin Recognition Sequences

Integrin	Interacting Sequences[a]	References
$\alpha_2\beta_1$	DGEA (Col1), RGD*[1]	10, 11
$\alpha_3\beta_1$	RGDS (Fn)	12
$\alpha_4\beta_1$	EILDVPS, REDV (Fn-alt), IDAPS (Fnhep), IDSP (VCAM-1), RGD[1]	13–16 17
$\alpha_5\beta_1$	RGDS (Fn), NGRAHA[1], RRETAWA*[1], PHSRN (Fn)§, KQAGDV[1]	18–20 21, 22
$\alpha_v\beta_1$	RGDV (Vn), RGDS (Fn)	23, 24
$\alpha_v\beta_3$	RGDV (Vn)	25
$\alpha_v\beta_5$	RGDV (Vn), RGDS (Fn), NGRAHA[1]	18, 26, 27
$\alpha_v\beta_6$	RGDS (Fn)	28
$\alpha_{IIb}\beta_3$	RGD (Fb#, vWF), KQAGDV (Fb), KGD (barbourin), DRVPHSRNSIT (Fn)	29, 30 31, 32
$\alpha_M\beta_2$	QKRLDGS (Fb)	33

[a] Integrin recognition sequences are summarized. Proteins from which these integrin-interacting sequences are derived are indicated in parentheses: Fn = fibronectin central cell-binding domain; Fn-alt = fibronectin alternatively spliced region; Fn-hep = fibronectin heparin binding domain; Vn = vitronectin; Col1 = collagen type I; vWF = von Willebrand factor. The # indicates that the sequence in the protein may not function in the native protein. Known minimal recognition sequences are marked by underlines. The * indicates that the sequence functions when it is circularized. A[1] indicates that synthetic peptides containing the sequence have been shown to posess inhibitory activities. A § indicates that the sequence functions when covalently located in cis to the RGD sequence.

Many of these new interacting sequences listed in Table I seem to bind to each integrin in a mutually competitive manner, suggesting that the binding site(s) on the integrin molecule are identical or extensively overlap for these sequences. For example, RGD-containing peptides can inhibit integrin $\alpha_4\beta_1$-mediated cell attachment to EILDV-conjugated substrates (15), and B lymphocytes can attach to the central cell-binding domain of fibronectin via activated $\alpha_4\beta_1$ in an RGD- and EILDV-dependent manner (37). The KQAGDV sequence in the fibrinogen γ chain acts as a closely related analog of RGD, because peptides containing this sequence interact with platelet integrin $\alpha_{IIb}\beta_3$ interchangeably with RGD-containing peptides. Although an aspartic acid residue has been postulated to be a key component of many integrin-binding sequences, a few sequences that do not have an aspartic acid have been shown to interact with integrins competitively with the RGD sequence. Although it is therefore very difficult to deduce a common feature of integrin-binding sequences at present, they seem to contain either a key arginine or an aspartic acid, or both.

Even though many proteins contain the various potential integrin interaction sequences discussed above, these potential sites are not always involved in cell adhesion. As expected in the cases of fibronectin, von Willebrand factor, and vitronectin, mutagenesis of their RGD sequences and expression of the mutagenized proteins results in decreased or complete loss of cell-adhesive activity (38–41), thereby supporting the results obtained from synthetic peptide studies. In contrast, however, mutations of the two RGD sequences in the fibrinogen α chain did not affect binding to integrin αIIbβ3 as measured by platelet aggregation assays. Nevertheless, a mutation of the γ chain peptide sequence KQAGDV (42) resulted in a large decrease of activity. The two RGD sequences on the fibrinogen α chain may therefore be hidden or not presented in a proper conformation. This characteristic of being cryptic may be also the case for some integrin-binding motifs in other extracellular matrix proteins, most notably the RGD sequence in the mouse laminin A chain (43), in human complement component C3 (44), and in the human immunodeficiency virus (HIV) tat protein (45).

As summarized briefly above, recent studies have identified many interaction motifs of integrins. However, there are still unanswered questions about the differences in specificity and affinity between synthetic peptides containing these binding motifs and native extracellular matrix proteins.

IV. Disintegrins and Cyclic Peptides

Disintegrins represent a family of snake venom proteins that have been shown to be very potent inhibitors of fibrinogen binding to the major platelet integrin αIIbβ3. Disintegrins are 100–1000 times more potent than linear RGD-containing peptides [for a review, see (46, 47)]. Many disintegrin family members also bind strongly to the integrins αvβ3 and α5β1 in an RGD-dependent manner. These proteins have a conserved RGD sequence, and mutagenesis studies have underscored the importance of the RGD sequence (48). One disintegrin termed barbourin has been identified as a very specific inhibitor of ligand binding to αIIbβ3, and it contains a KGD sequence instead of RGD (31). This conservative substitution appears to be a key feature of its integrin specificity (for interactions with αIIbβ3, but not with αvβ3 and α5β1), since R to K substitution mutants of other disintegrins have been shown to bind to αIIbβ3 at levels similar to wild-type proteins, but not to bind to αvβ3 and α5β1 (31, 49).

There are a variety of reports on disintegrin-like proteins that are potent antagonists of the integrin αIIbβ3, including mambin (50) and the leech proteins decorsin (51) and ornatin (52), which have 3-D structures maintained by disulfide bonds and a RGD sequence. However, the disintegrin-like protein termed disagregin isolated from ticks (35) is also an effective antagonist, even though it contains neither an RGD nor a KGD sequence.

Conformationally constraining RGD peptides by circularization (i.e., formation of cyclic peptides) has been shown to produce inhibitors of ligand binding that have much higher affinity and specificity for αIIbβ3 and αvβ3, but not α5β1, compared to the corresponding linear peptides (53). There are now many extensive studies on the effects of conformational constraint of integrin-binding sequences, on the effects of amino acid sequences adjacent to the RGD sequence in circularized peptides, and on the effects of modifications of RGD (54). These studies generally agree with the conclusion that circularization of RGD-containing peptides (or a related sequence) and placement of a hydrophobic amino acid residue at the X position of RGDX combine to enhance binding to the integrin αIIbβ3. Although determination of the 3-D structure of an integrin will be necessary to understand this specificity in detail, integrin αIIbβ3 has been suggested to be able to accommodate a larger distance between arginine and aspartic acid side chains than integrin αvβ3 and α5β1 (55).

The high affinity of disintegrins to αIIbβ3 is likely to arise from the conformational constraint of their RGD sequence at the tip of an arm consisting of two antiparallel polypeptide chains, as shown by the 3-D structure determinations of several disintegrins and decorsin (56–61). Detailed alanine-scanning mutagenesis of kistrin also indicates the primary importance of the RGD sequence (47). Some unanswered questions remain, however: (a) high-affinity binding to integrin α5β1 of some disintegrins has not been reproduced by the corresponding circularized peptides, and (b) the RGD sequences on the disintegrins except decorsin (61) seem to have more relaxed conformations (less conformational constraint) than cyclic peptides.

V. Central Cell-Binding Domain of Fibronectin

Figure 2 shows a schematic representation of fibronectin structure and several conceptually informative proteolytic or recombinant fragments from the cell-binding domain. Historically, proteolytic fragments have been used for mapping the cell-adhesive site of fibronectin. Among many proteolytic fragments, the 11,500-Da pepsin fragment containing RGD (62) and corresponding approximately to the 10th type III repeating unit, was the smallest, but not fully active cell-adhesive fragment. Although the latter peptide retains activity for interaction with vitronectin receptors, it has 50–200-fold less adhesive activity and affinity for the $\alpha 5 \beta 1$ integrin (22, 63–65). In contrast, a longer fragment (75,000 Da) has much higher and more specific activity, comparable to that of intact fibronectin (63).

It is difficult to determine the minimal fully active cell-binding domain using proteolytic fragments. Even after several expression systems for truncated fibronectin fragments were established, there were conflicting reports concerning the determination of the minimum, fully active cell-binding domain of fibronectin (39, 64, 66). A recent study indicates that this confusion was caused by artifactual effects resulting from adsorbing short polypeptides directly to plastic surfaces. In fact, in striking contrast to the 11,500-Da fragment, a short 20,000-Da fragment containing the 9th and 10th fibronectin repeating units (Fig. 2) has shown nearly full cell-adhesive activity when this protein fragment is assayed in solution by competitive inhibition assays, or after binding above a plastic surface by using a noninhibiting monoclonal antibody to "present" it appropriately to cells (65). The 20,000-Da fragment appear to represent the smallest fully active segment of the cell-binding domain of fibronectin (Fig. 2).

Detailed mutagenesis studies (21, 66) suggested that the central region of the 9th type III repeat (a "synergy" site with amino acid sequence PHSRN) is important for full activity of the cell-binding domain of fibronectin for cell adhesion dependent on the integrin $\alpha 5 \beta 1$ (Fig. 3). Amino acid substitution studies within this synergy sequence indicate that the sequence specificity of the PHSRN se-

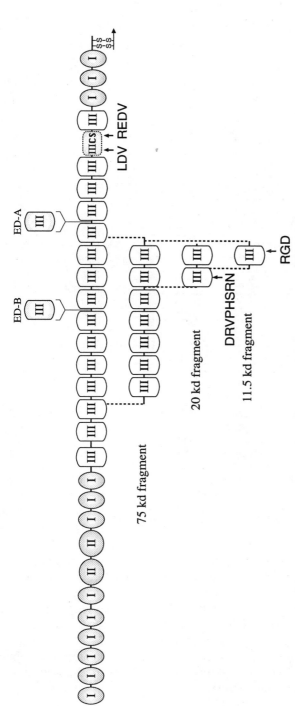

Figure 2. Schematic structure of fibronectin. The fibronectin molecule contains three types of basic repeating unit, termed types I, II, and III. Peptide adhesive interaction sequences are listed, and several biologically active fragments that have been characterized in detail are shown.

9

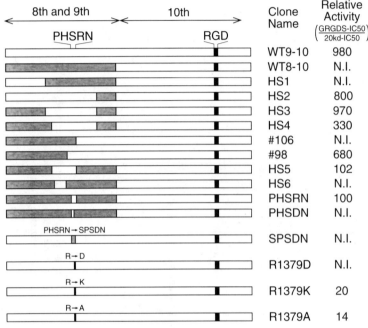

Figure 3.   Homology scanning within the 9th type III repeat. The various recombination mutants shown were constructed and assayed by a competitive inhibition assay for cell adhesion to fibronectin. Amino acid residues from the 8th repeat are indicated by gray boxes. After replacing various parts of the inactive clone WT8-10 with sequences from the 9th repeat, inhibitory activity was measured (21). The concentrations required for 50% inhibition of BHK cell spreading are indicated relative to activity of the GRGDS synthetic peptide (IC50 of the GRGDS peptide/IC50 of the protein). "N.I." (no inhibition detected) indicates that no inhibitory activity whatsoever was detected at the highest concentration tested.

quence is quite low, but that its arginine residue is the most important for interaction with $\alpha5\beta1$. Interestingly, a synthetic peptide containing this sequence can enhance the cell-adhesive activity of RGD for $\alpha5\beta1$ only when it is covalently linked to the corresponding region of a nonactive construct (21). This result suggests that the sequence is effective with this integrin only when it is located at a certain position cis to the RGD sequence. A monoclonal antibody that binds to this sequence has been found to inhibit cell adhesion, migration, and matrix assembly (21, 65).

In an independent study, the binding of the major platelet integrin αIIbβ3 to this region has been examined (32, 67). This integrin has also been shown to bind to the central region of the 9th repeat. However, unlike integrin α5β1-dependent cell adhesion, a synthetic peptide from this region (DRVPHSRNSIT) exhibits significant inhibitory activity in an αIIbβ3-dependent platelet aggregation assay, without any requirement for covalent positioning. The reason for this difference is not yet clear, but platelet aggregation assays may be more sensitive than cell-spreading assays, or these peptides may function by somewhat different molecular mechanisms. In fact, an amino acid substitution study of the DRVPHSRNSIT peptide has indicated that the first aspartic acid and the second arginine residues preceding the PHSRN sequence are especially important for inhibition of integrin-mediated platelet aggregation; in contrast, replacement of the arginine residue at position 7 in a synthetic peptide corresponding to this site resulted in minimal loss of activity for αIIbβ3 (32), a finding that differs from conclusions from the mutagenesis analyses of the requirements for interactions with α5β1. Thus, even though the α5β1 and αIIbβ3 integrins appear to utilize similar synergy or enhancing sequences, they may differ in the amino acids most important for activity and in the need for positioning relative to the RGD sequence.

The fibronectin type III repeating unit is a protein motif found in many proteins including growth hormone receptors and other extracellular matrix proteins, such as tenascin. The 3-D structure of this motif has been determined in several proteins: human growth hormone receptor extracellular domain-growth hormone complex (68), one type III repeating unit (10th) of fibronectin (69, 70), one of the repeating units of tenascin (71), and 2 units from *Drosophila* neuroglian (72) have been determined. All structural data indicate a barrel-shaped structure consisting of two β sheets facing each other.

These studies and molecular modeling suggest that the 3-D structure of the minimal fully active cell-binding domain (the 20,000-Da fragment discussed above) consists of two barrel-shaped type III repeating units rotated 180° against each other (Fig. 4). Importantly, the RGD sequence is located in a relatively flexible loop (69). In this model, the PHSRN sequence is interestingly also located in a loop, but which is unlikely to have direct contact with the RGD sequence. Although we must await direct determination of the 3-D structure

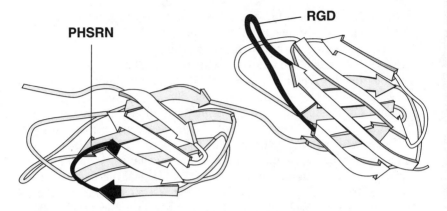

Figure 4. A model for the 3-D structure of the cell-binding domain of fibronectin. This model was constructed based on the published structure of single FN type III repeats (see text). In this model, the 9th repeat is rotated by 180° relative to the 10th repeat unit. The RGD and the PHSRN loops are indicated in black.

of a fully active cell-binding domain fragment, which is underway in several laboratories, it appears likely that a second site besides the RGD sequence rather than a special constrained conformation of the RGD sequence is necessary for full activity during interaction of fibronectin with the α5β1 integrin.

VI.   Fibronectin IIICS Region

The other major cell-adhesive domain in fibronectin is termed the IIICS (type III connecting sequence) region. It has several forms that originate from differential precursor mRNA splicing (Fig. 5), and it has no sequence similarity with the type III repeating motif. The Glu-Ile-Leu-Asp-Val-Pro-Ser (EILDVPS) sequence is located in the CS1 region, and has been shown to bind to integrin α4β1 (73, 74). Another sequence, Arg-Glu-Asp-Val (REDV) in the CS5 region, has been also shown to have binding activity for α4β1 (15). A third sequence, Ile-Asp-Ala-Pro-Ser (IDAPS) located near the carboxyl terminus of the heparin-binding domain of fibronectin, is also recognized by the integrin α4β1 (16, 75). These three peptides, RGD-

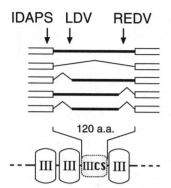

Figure 5. Schematic structure of the fibronectin IIICS region. There are at least five alternatively spliced forms in the type III connecting sequence (IIICS) region of fibronectin (1). The three binding sites of the integrin α4β1 are indicated by arrows; a.a., amino acids.

containing peptides (15), and VCAM-1 (76) all seem to compete for an identical site or spacially overlapping sites on the integrin α4β1. Because the minimal essential sequence of the CS1 peptide sequence was determined as LDV, the binding sequence motif for integrin α4β1 may be merely X-Asp-Y, where X (Gly, Leu, or Glu) and Y (Ser or Val) is tolerated (14, 17). Although the central cell-binding domain of fibronectin can promote cell attachment of many cell types, the IIICS region can also mediate cell adhesion and migration of some biologically interesting cell types, such as neural crest cells and their derivatives and lymphocytes (77, 78).

It has been reported that long peptides from CS-1 show relatively high activity, but shorter peptides gradually lose activity (14). Because five proline residues as well as the LDV sequence are conserved from chicken to human in the CS1 region, the conformational presentation of the LDV sequence or additional specific amino acid sequences may play important roles in the affinity and specificity of integrin α4β1 interactions with the CS-1 region. It is also still unclear whether the three integrin α4β1-binding sites in this region of fibronectin function independently or interactively with each other.

VII. Discussion

Recent studies have identified an increasing number of integrin interaction sites present in fibronectin and other extracellular matrix proteins. These recognition sequences are usually short peptide motifs, and several such sequences seem to compete for interaction with overlapping or identical functional-binding sites in an integrin. This characteristic may be explained by the evolution of integrins from a common ancestral molecule that had a binding site for short amino acid sequence(s).

Current data support a model in which the ligand-binding site(s) on integrins consist of two regions near the amino-terminal ends of both α and β subunits. Extensive studies using the αIIbβ3 platelet integrin suggest the presence of a complex, 3-D binding pocket at the interface of the two subunits (79, 80). Most of the identified recognition sequences probably interact with this pocket competitively, though the possibility of an allosteric-binding site cannot be excluded. Although there have been some reports that suggest the presence of other ligand-binding site(s) on integrins, they have not yet been characterized in detail.

The simple presence of a cell-adhesion motif in a protein is not sufficient to predict adhesive function. In fact, some motifs such as RGD that are present in even some cell-adhesive proteins, such as laminin, may not normally function in the native protein because the site is hidden or is not in a proper conformation. Some sites can be exposed by denaturation or possibly by proteolytic degradation of the original extracellular molecule. As expected, the active sites whose structures have been determined so far (fibronectin and disintegrins) are readily accessible from the outside of the protein and are usually located in a relatively flexible loop. In the future, determination of the 3-D molecular structures of integrins, as well as those of other cell-adhesive proteins will provide valuable information about the physical basis of ligand–integrin interactions and of specificity.

Determining the molecular basis for the differences in affinity and specificity between short peptide recognition sequences and native integrin ligands is one of the most interesting research topics in the field of integrin–ligand interaction. Conformational effect(s) as well as adjacent amino acid sequences certainly affect the specificity and affinity of ligand binding to integrins. The platelet integrin αIIbβ3 is the most extensively characterized example. For some integrin ligands, conformation is known to be particularly important; for example, in disintegrins, the RGD motif is located at the tip of a long arm formed by two antiparallel polypeptide strands. Linearization of disintegrin molecules by breaking their disulfide bonds has been shown to destroy their activity. Invasin, which binds strongly to some integrins, also requires structure maintained by disulfide bonds (81).

However, there is increasing evidence that secondary information besides the short integrin recognition sequences is necessary for full activity of some cell-adhesive proteins. At least in the case of the cell-binding domain of fibronectin, the conformation of the RGD sequence alone does not seem to explain the difference in cell-adhesive activity between the longer fragments and RGD-containing peptides. Current studies and molecular modeling suggest that the conformation of the 10th type III repeat of fibronectin, which has the RGD sequence, may be similar in the 11,500-Da fragment and in longer fragments, although final determination of the 3-D molecular structure of a longer fragment will be necessary. Detailed studies of this region support the notion that certain integrins can use a secondary "synergy" site that enhances the cell-adhesive activity of the RGD sequence when it is positioned at a certain distance from the RGD sequence. Future investigations should determine whether this feature is common among other integrins. The molecular mechanism of this enhancement is also still unknown, although the secondary site is likely to interact directly with integrins. Because some disintegrins and invasin can bind strongly even to integrins that do not appear to bind preferentially to conformationally stable, circularized binding sequences, these protein ligands may also have a secondary site(s) for high-affinity interaction with these integrins.

Fibronectin has served as a valuable model system for analyzing cell adhesion for many years. There are still unanswered questions, for example, about the molecular mechanisms of its interactions and the coordination between its different adhesive domains. Further studies of integrin–fibronectin interactions are important, especially considering the wide variety of biological processes in which integrin–ligand interactions play important roles. Findings from these studies may lead to the development of highly active and specific artificial integrin ligands, which may be immediately applicable to various medical and therapeutic purposes, for example, in the areas of wound repair, control of platelet function, and tumor cell metastasis.

References

1. Kornblihtt, A. R., Umezawa, K., Vibe, P. K., and Baralle, F. E., Primary structure of human fibronectin: differential splicing may generate at least 10 polypeptides from a single gene, *EMBO J.*, **4**, 1755–1759 (1985).

2. Akiyama, S. K. and Yamada, K. M., Fibronectin, *Adv. Enzymol.*, **59**, 1–57 (1987).

3. Mosher, D. F., *Fibronectin*, Academic Press, New York, 1989.

4. Hynes, R. O., *Fibronectins*, Springer-Verlag, New York, 1990.

5. Yamada, K. M., Adhesive recognition sequences, *J. Biol. Chem.*, **266**, 12809–12812 (1991).

6. D'Souza, S. E., Ginsberg, M. H., and Plow, E. F., Arginyl-glycyl-aspartic acid (RGD): a cell adhesion motif, *Trends Biochem. Sci.*, **16**, 246–250 (1991).

7. Hynes, R. O., Integrins: versatility, modulation, and signaling in cell adhesion, *Cell*, **69**, 11–25 (1992).

8. Sonnenberg, A., Integrins and their ligands, *Curr. Top. Microbiol. Immunol.*, **184**, 7–35 (1993).

9. Ruoslahti, E., Noble, N. A., Kagami, S., and Border, W. A., Integrins, *Kidney Int.*, **45**, S17–S22 (1994).

10. Staatz, W. D., Fok, K. F., Zutter, M. M., Adams, S. P., Rodriguez, B. A., and Santoro, S. A., Identification of a tetrapeptide recognition sequence for the $\alpha2\beta1$ integrin in collagen, *J. Biol. Chem.*, **266**, 7363–7367 (1991).

11. Cardarelli, P. M., Yamagata, S., Taguchi, I., Gorcsan, F., Chiang, S. L., and Lobl, T., The collagen receptor $\alpha2\beta1$, from MG-63 and HT1080 cells, interacts with a cyclic RGD peptide, *J. Biol. Chem.*, **267**, 23159–23164 (1992).

12. Elices, M. J., Urry, L. A., and Hemler, M. E., Receptor functions for the integrin VLA-3: fibronectin, collagen, and laminin binding are differentially influenced by Arg-Gly-Asp peptide and by divalent cations, *J. Cell Biol.*, **112**, 169–181 (1991).

13. Humphries, M. J., Komoriya, A., Akiyama, S. K., Olden, K., and Yamada, K. M., Identification of two distinct regions of the type III connecting segment of human plasma fibronectin that promote cell type-specific adhesion, *J. Biol. Chem.*, **262**, 6886–6892 (1987).

14. Komoriya, A., Green, L. J., Mervic, M., Yamada, S. S., Yamada, K. M., and Humphries, M. J., The minimal essential sequence for a major cell type-specific adhesion site (CS1) within the alternatively spliced type III connecting segment domain of fibronectin is leucine-aspartic acid-valine, *J. Biol. Chem.*, **266**, 15075–15079 (1991).

15. Mould, A. P., Komoriya, A., Yamada, K. M., and Humphries, M. J., The CS5 peptide is a second site in the IIICS region of fibronectin recognized by the integrin $\alpha4\beta1$. Inhibition of $\alpha4\beta1$ function by RGD peptide homologues, *J. Biol. Chem.*, **266**, 3579–3585 (1991).

16. Mould, A. P. and Humphries, M. J., Identification of a novel recognition sequence for the integrin $\alpha4\beta1$ in the COOH-terminal heparin-binding domain of fibronectin, *EMBO J.*, **10**, 4089–4095 (1991).

17. Clements, J. M., Newham, P., Shepherd, M., Gilbert, R., Dudgeon, T. J., Needham, L. A., Edwards, R. M., Berry, L., Brass, A., and Humphries, M. J., Identification of a key integrin-binding sequence in VCAM-1 homologous to the LDV active-site in fibronectin, *J. Cell Sci.,* **107,** 2127–2135 (1994).

18. Koivunen, E., Gay, D. A., and Ruoslahti, E., Selection of peptides binding to the α5β1 integrin from phage display library, *J. Biol. Chem.,* **268,** 20205–20210 (1993).

19. Pierschbacher, M. D. and Ruoslahti, E., Cell attachment activity of fibronectin can be duplicated by small synthetic fragments of the molecule, *Nature,* **309,** 30–33 (1984).

20. Koivunen, E., Wang, B., and Ruoslahti, E., Isolation of a highly specific ligand for the α5β1 integrin from a phage display library, *J. Cell Biol.,* **124,** 373–380 (1994).

21. Aota, S., Nomizu, M., and Yamada, K. M., The short amino acid sequence Pro-His Ser-Arg-Asn in human fibronectin enhances cell-adhesive function, *J. Biol. Chem.,* **269,** 24756–24761 (1994).

22. Hautanen, A., Gailit, J., Mann, D. M., and Ruoslahti, E., Effects of modifications of the RGD sequence and its context on recognition by the fibronectin receptor, *J. Biol. Chem.,* **264,** 1437–1442 (1989).

23. Vogel, B. E., Tarone, G., Giancotti, F. G., Gailit, J., and Ruoslahti, E., A novel fibronectin receptor with an unexpected subunit composition (αvβ1), *J. Biol. Chem.,* **265,** 5934–5937 (1990).

24. Zhang, Z., Morla, A. O., Vuori, K., Bauer, J. S., Juliano, R. L., and Ruoslahti, E., The αvβ1 integrin functions as a fibronectin receptor but does not support fibronectin matrix assembly and cell migration on fibronectin, *J Cell Biol,* **122,** 235–242 (1993).

25. Pytela, R., Pierschbacher, M. D., and Ruoslahti, E., A 125/115-kDa cell surface receptor specific for vitronectin interacts with the arginine-glycine-aspartic acid adhesion sequence derived from fibronectin, *Proc. Natl. Acad. Sci. USA,* **82,** 5766–5770 (1985).

26. Smith, J. W., Vestal, D. J., Irwin, S. V., Burke, T. A., and Cheresh, D. A., Purification and functional characterization of integrin αvβ5. An adhesion receptor for vitronectin, *J. Biol. Chem.,* **265,** 11008–11013 (1990).

27. Pasqualini, R., Bodorova, J., Ye, S., and Hemler, M. E., A study of the structure, function and distribution of β5 integrins using novel anti-β5 monoclonal antibodies, *J. Cell Sci.,* **105,** 101–111 (1993).

28. Busk, M., Pytela, R., and Sheppard, D., Characterization of the integrin αvβ6 as a fibronectin-binding protein, *J. Biol. Chem.,* **267,** 5790–5796 (1992).

29. Kloczewiak, M., Timmons, S., Lukas, T. J., and Hawiger, J., Platelet receptor recognition site on human fibrinogen. Synthesis and structure-function relationship of peptides corresponding to the carboxy-terminal segment of the γ chain, *Biochemistry,* **23,** 1767–1774 (1984).

30. Plow, E. F., Pierschbacher, M. D., Ruoslahti, E., Marguerie, G., and Ginsberg, M. H., Arginyl-glycyl-aspartic acid sequences and fibrinogen binding to platelets, *Blood,* **70,** 110–115 (1987).

31. Scarborough, R. M., Rose, J. W., Hsu, M. A., Phillips, D. R., Fried, V. A., Campbell, A. M., Nannizzi, L., and Charo, I. F., Barbourin. A GPIIb-IIIa-specific integrin antagonist from the venom of *Sistrurus m. barbouri, J. Biol. Chem.,* **266,** 9359–9362 (1991).

32. Bowditch, R. D., Hariharan, M., Tominna, E. F., Smith, J. W., Yamada, K. M., Getzoff, E. D., and Ginsberg, M. H., Identification of a novel integrin binding site in fibronectin. Differential utilization by β3 integrins, *J. Biol. Chem.,* **269,** 10856–10863 (1994).

33. Altieri, D. C., Agbanyo, F. R., Plescia, J., Ginsberg, M. H., Edgington, T. S., and Plow, E. F., A unique recognition site mediates the interaction of fibrinogen with the leukocyte integrin Mac-1 (CD11b/CD18), *J. Biol. Chem.,* **265,** 12119–12122 (1990).

34. Isberg, R. R. and Leong, J. M., Multiple β1 chain integrins are receptors for invasin, a protein that promotes bacterial penetration into mammalian cells, *Cell,* **60,** 861–871 (1990).

35. Karczewski, J., Endris, R., and Connolly, T. M., Disagregin is a fibrinogen receptor antagonist lacking the Arg-Gly-Asp sequence from the tick, *Ornithodoros moubata, J. Biol. Chem.,* **269,** 6702–6708 (1994).

36. O'Neil, K. T., Hoess, R. H., Jackson, S. A., Ramachandran, N. S., Mousa, S. A., and DeGrado, W. F., Identification of novel peptide antagonists for GPIIb/IIIa from a conformationally constrained phage peptide library, *Proteins,* **14,** 509–515 (1992).

37. Sanchez-Aparicio, P., Dominguez-Jimenez, C., and Garchia-Pardo, A., Activation of the α4β1 integrin through the β1 subunit induces recognition of the RGDS sequence in fibronectin., *J. Cell Biol.,* **126,** 271–279 (1994).

38. Obara, M., Kang, M. S., and Yamada, K. M., Site-directed mutagenesis of the cell-binding domain of human fibronectin: separable, synergistic sites mediate adhesive function, *Cell,* **53,** 649–657 (1988).

39. Kimizuka, F., Ohdate, Y., Kawase, Y., Shimojo, T., Taguchi, Y., Hashino, K., Goto, S., Hashi, H., Kato, I., Sekiguchi, K., and Titani, K., Role of type III homology repeats in cell adhesive function within the cell-binding domain of fibronectin, *J. Biol. Chem.,* **266,** 3045–3051 (1991).

40. Beacham, D. A., Wise, R. J., Turci, S. M., and Handin, R. I., Selective inactivation of the Arg-Gly-Asp-Ser (RGDS) binding site in von Willebrand factor by site-directed mutagenesis, *J. Biol. Chem.,* **267,** 3409–3415 (1992).

41. Cherny, R. C., Honan, M. A., and Thiagarajan, P., Site-directed mutagenesis of the arginine-glycine-aspartic acid in vitronectin abolishes cell adhesion, *J. Biol. Chem.,* **268,** 9725–9729 (1993).

42. Farrell, D. H., Thiagarajan, P., Chung, D. W., and Davie, E. W., Role of fibrinogen α and γ chain sites in platelet aggregation, *Proc. Natl. Acad. Sci. USA,* **89,** 10729–10732 (1992).

43. Aumailley, M., Gerl, M., Sonnenberg, A., Deutzmann, R., and Timpl, R., Identification of the Arg-Gly-Asp sequence in laminin A chain as a latent cell-binding site being exposed in fragment P1, *FEBS Lett.,* **262,** 82–86 (1990).

44. Taniguchi-Sidle, A., and Isenman, D. E., Mutagenesis of the Arg-Gly-Asp triplet in human complement component C3 does not abolish binding of iC3b to the leukocyte integrin complement receptor type III (CR3, CD11b/CD18), *J. Biol. Chem.*, **267**, 635–643 (1992).

45. Vogel, B. E., Lee, S. J., Hildebrand, A., Craig, W., Pierschbacher, M. D., Wong, S. F., and Ruoslahti, E., A novel integrin specificity exemplified by binding of the $\alpha v \beta 5$ integrin to the basic domain of the HIV Tat protein and vitronectin, *J. Cell Biol.*, **121**, 461–468 (1993).

46. Blobel, C. P. and White, J. M., Structure, function and evolutionary relationship of proteins containing a disintegrin domain, *Curr. Opin. Cell Biol.*, **4**, 760–765 (1992).

47. Lazarus, R. A. and McDowell, R. S., Structural and functional aspects of RGD-containing protein antagonists of glycoprotein IIb-IIIa, *Curr. Opin. Biotech.*, **4**, 438–445 (1993).

48. Dennis, M. S., Carter, P., and Lazarus, R. A., Binding interactions of kistrin with platelet glycoprotein IIb-IIIa: analysis by site-directed mutagenesis, *Proteins*, **15**, 312–321 (1993).

49. Scarborough, R. M., Naughton, M. A., Teng, W., Rose, J. W., Phillips, D. R., Nannizzi, L., Arfsten, A., Campbell, A. M., and Charo, I. F., Design of potent and specific integrin antagonists. Peptide antagonists with high specificity for glycoprotein IIb-IIIa, *J. Biol. Chem.*, **268**, 1066–1073 (1993).

50. McDowell, R. S., Dennis, M. S., Louie, A., Shuster, M., Mulkerrin, M. G., and Lazarus, R. A., Mambin, a potent glycoprotein IIb-IIIa antagonist and platelet aggregation inhibitor structurally related to the short neurotoxins, *Biochemistry*, **31**, 4766–4772 (1992).

51. Seymour, J. L., Henzel, W. J., Nevins, B., Stults, J. T., and Lazarus, R. A., Decorsin. A potent glycoprotein IIb-IIIa antagonist and platelet aggregation inhibitor from the leech *Macrobdella decora*, *J. Biol. Chem.*, **265**, 10143–10147 (1990).

52. Mazur, P., Henzel, W. J., Seymour, J. L., and Lazarus, R. A., Ornatins: potent glycoprotein IIb-IIIa antagonists and platelet aggregation inhibitors from the leech *Placobdella ornata*, *Eur. J. Biochem.*, **202**, 1073–1082 (1991).

53. Pierschbacher, M. D. and Ruoslahti, E., Influence of stereochemistry of the sequence Arg-Gly-Asp-Xaa on binding specificity in cell adhesion, *J. Biol. Chem.*, **262**, 17294–17298 (1987).

54. Blackburn, B. K. and Gadek, T. R., Glycoprotein IIbIIIa antagonists, *Ann. Rep. Med. Chem.*, **28**, 79–88 (1993).

55. Pfaff, M., Tangemann, K., Muller, B., Gurrath, M., Muller, G., Kessler, H., Timpl, R., and Engel, J., Selective recognition of cyclic RGD peptides of NMR defined conformation by $\alpha IIb \beta 3$, $\alpha v \beta 3$, and $\alpha 5 \beta 1$ integrins, *J. Biol. Chem.*, **269**, 20233–20238 (1994).

56. Adler, M., Lazarus, R. A., Dennis, M. S., and Wagner, G., Solution structure of kistrin, a potent platelet aggregation inhibitor and GP IIb-IIIa antagonist, *Science*, **253**, 445–448 (1991).

57. Saudek, V., Atkinson, R. A., Lepage, P., and Pelton, J. T., The secondary structure of echistatin from *1H*-NMR, circular-dichroism and Raman spectroscopy, *Eur. J. Biochem.*, **202**, 329–338 (1991).

58. Cooke, R. M., Carter, B. G., Martin, D. M., Murray, R. P., and Weir, M. P., Nuclear magnetic resonance studies of the snake toxin echistatin. 1H resonance assignments and secondary structure, *Eur. J. Biochem.*, **202**, 323–328 (1991).

59. Dalvit, C., Widmer, H., Bovermann, G., Breckenridge, R., and Metternich, R., 1H NMR studies of echistatin in solution. Sequential resonance assignments and secondary structure, *Eur. J. Biochem.*, **202**, 315–321 (1991).

60. Senn, H. and Klaus, W., The nuclear magnetic resonance solution structure of flavoridin, an antagonist of the platelet GP IIb-IIIa receptor, *J. Mol. Biol.*, **232**, 907–925 (1993).

61. Krezel, A. M., Wagner, G., Seymour, U. J., and Lazarus, R. A., Structure of the RGD protein decorsin: conserved motif and distinct function in leech proteins that affect blood clotting, *Science,* **264**, 1944–1947 (1994).

62. Pierschbacher, M. D., Hayman, E. G., and Ruoslahti, E., Location of the cell-attachment site in fibronectin with monoclonal antibodies and proteolytic fragments of the molecule., *Cell,* **26**, 259–267 (1981).

63. Akiyama, S. K., Hasegawa, E., Hasegawa, T., and Yamada, K. M., The interaction of fibronectin fragments with fibroblastic cells, *J. Biol. Chem.*, **260**, 13256–13260 (1985).

64. Nagai, T., Yamakawa, N., Aota, S., Yamada, S. S., Akiyama, S. K., Olden, K., and Yamada, K. M., Monoclonal antibody characterization of two distant sites required for function of the central cell-binding domain of fibronectin in cell adhesion, cell migration, and matrix assembly, *J. Cell Biol.*, **114**, 1295–305 (1991).

65. Akiyama, S. K., Aota, S., and Yamada, K. M., Function and receptor specificity of a minimal 20 kilodalton cell adhesive fragment of fibronectin, *Cell Adh. Commun.*, **3**, 13–25 (1994).

66. Aota, S., Nagai, T., and Yamada, K. M., Characterization of regions of fibronectin besides the arginine-glycine-aspartic acid sequence required for adhesive function of the cell-binding domain using site-directed mutagenesis, *J. Biol. Chem.*, **266**, 15938–15943 (1991).

67. Bowditch, R. D., Halloran, C. E., Aota, S., Obara, M., Plow, E. F., Yamada, K. M., and Ginsberg, M. H., Integrin αIIbβ3 (platelet GPIIb-IIIa) recognizes multiple sites in fibronectin, *J. Biol. Chem.*, **266**, 23323–23328 (1991).

68. deVos, A. M., Ultsch, M., and Kossiakoff, A. A., Human growth hormone and extracellular domain of its receptor: crystal structure of the complex, *Science,* **255**, 306–312 (1992).

69. Main, A. L., Harvey, T. S., Baron, M., Boyd, J., and Campbell, I. D., The three-dimensional structure of the tenth type III module of fibronectin: an insight into RGD-mediated interactions, *Cell,* **71**, 671–678 (1992).

70. Dickinson, C. D., Veerapandian, B., Dai, X. P., Hamlin, R. C., Xuong, N. H., Ruoslahti, E., and Ely, K. R., Crystal structure of the tenth type III cell adhesion module of human fibronectin, *J. Mol. Biol.*, **236**, 1079–1092 (1994).

71. Leahy, D. J., Hendrickson, W. A., Aukhil, I., and Erickson, H. P., Structure of a fibronectin type III domain from tenascin phased by MAD analysis of the selenomethionyl protein, *Science,* **258,** 987–991 (1992).

72. Huber, A. H., Wang, Y. M., Bieber, A. J., and Bjorkman, P. J., Crystal structure of tandem type III fibronectin domains from Drosophila neuroglian at 2.0 A, *Neuron,* **12,** 717–731 (1994).

73. Mould, A. P., Wheldon, L. A., Komoriya, A., Wayner, E. A., Yamada, K. M., and Humphries, M. J., Affinity chromatographic isolation of the melanoma adhesion receptor for the IIICS region of fibronectin and its identification as the integrin α4β1, *J. Biol. Chem.,* **265,** 4020–4024 (1990).

74. Guan, J. L., and Hynes, R. O., Lymphoid cells recognize an alternatively spliced segment of fibronectin via the integrin receptor α4β1, *Cell,* **60,** 53–61 (1990).

75. Wayner, E. A., Garcia, P. A., Humphries, M. J., McDonald, J. A., and Carter, W. G., Identification and characterization of the T lymphocyte adhesion receptor for an alternative cell attachment domain (CS-1) in plasma fibronectin, *J. Cell Biol.,* **109,** 1321–1330 (1989).

76. Makarem, R., Newham, P., Askari, J. A., Green, L. J., Clements, J., Edwards, M., Humphries, M. J., and Mould, A. P., Competitive binding of vascular cell adhesion molecule-1 and the HepII/IIICS domain of fibronectin to the integrin α4β1, *J. Biol. Chem.,* **269,** 4005–4011 (1994).

77. Dufour, S., Duband, J. L., Humphries, M. J., Obara, M., Yamada, K. M., and Thiery, J. P., Attachment, spreading and locomotion of avian neural crest cells are mediated by multiple adhesion sites on fibronectin molecules, *EMBO J.,* **7,** 2661–2671 (1988).

78. Garcia-Pardo, A., Wayner, E. A., Carter, W. G., and Ferreira, O. J., Human B lymphocytes define an alternative mechanism of adhesion to fibronectin. The interaction of the α4β1 integrin with the LHGPEILDVPST sequence of the type III connecting segment is sufficient to promote cell attachment, *J. Immunol.,* **144,** 3361–3366 (1990).

79. Plow, E. F., D'Souza, S. E., and Ginsberg, M. H., Ligand binding to GPIIb-IIIa: a status report, *Semin. Thromb. Hemost.,* **18,** 324–332 (1992).

80. Calvete, J. J., Mann, K., Schafer, W., Fernandez-Lafuente, R., and Guisan, J. M., Proteolytic degradation of the RGD-binding and non-RGD-binding conformers of human platelet integrin glycoprotein IIb/IIIa: clues for identification of regions involved in the receptor's activation, *Biochem. J.,* **298,** 1–7 (1994).

81. Leong, J. M., Morrissey, P. E., and Isberg, R. R., A 76-amino acid disulfide loop in the *Yersinia pseudotuberculosis* invasin protein is required for integrin receptor recognition, *J. Biol. Chem.,* **268,** 20524–20532 (1993).

EUKARYOTIC DIHYDROFOLATE REDUCTASE

By RAYMOND L. BLAKLEY, *Department of Molecular Pharmacology, St. Jude Children's Research Hospital, Memphis, Tennessee*

CONTENTS

Advances in Enzymology and Related Areas of Molecular Biology, Volume 70, Edited by Alton Meister.
ISBN 0-471-04097-5   © 1995 John Wiley & Sons, Inc.

I.   Introduction

Dihydrofolate reductase (DHFR) catalyses the reduction of 7,8-dihydrofolate by NADPH:

$$7,8\text{-Dihydrofolate} + \text{NADPH} + \text{H}^+$$

$$\rightarrow 5,6,7,8\text{-tetrahydrofolate} + \text{NADP}^+$$

The product, H$_4$folate, is essential for the synthesis of thymidylate, the synthesis of purine nucleotides, and other metabolic functions. Because of their greater accessibility, the bacterial DHFRs were investigated in detail prior to comparable studies on the eukaryotic enzymes. However, the advent of recombinant DNA technology has made it possible to study DHFR from many organisms with almost equal facility. As a consequence, eukaryotic DHFRs have received increasing attention in recent years.

The human enzyme is of particular interest because it is the target of cancer chemotherapy with methotrexate (MTX) and some related inhibitors, and has therefore been the subject of many recent studies. Despite the large body of information gathered on the prokaryotic DHFRs, direct studies of the vertebrate enzymes are essential for the detailed understanding of them that is required for the design of new inhibitors of DHFR for use as therapeutic agents, and for advances in the therapeutic use of such agents.

In this chapter the present state of knowledge of the eukaryotic DHFRs is summarized. The majority of this chapter deals with the vertebrate enzymes, simply because much more information is available about them compared with other eukaryotic DHFRs. We have not included the DHFRs from protozoa, in part because, in contrast to all other DHFRs, they are bifunctional enzymes in which DHFR is fused with thymidylate synthase. The primary structure of the DHFR portion of these proteins also contains large insertions compared with the sequence of other known eukaryotic DHFRs.

II. Structure

A. PRIMARY STRUCTURE

The sequences for six vertebrate DHFRs, two yeast DHFRs, and one fungal DHFR are given in Table I. The mammalian DHFRs have 186 residues, chicken DHFR 189, *Saccharomyces cerevisiae* DHFR 211, *Cryptococcus neoformans* DHFR 229, and *Pneumocystis carinii* DHFR 206. Aligning the sequences of the vertebrate DHFRs is straightforward since 130 residues (70%) are identical in all six sequences. Until crystal structures are available for the other eukaryotic DHFRs aligning their sequences with the vertebrate sequences is a matter of conjecture, certainly beyond the first 80 residues, since only 37 residues (20% of the hDHFR residues) are identical in all nine sequences according to the alignment in Table I. Even in the case of the two yeast sequences, only 64 residues are identical.

The extent of sequence identity within the vertebrate DHFRs is shown in Table II. It may be seen that the hamster, human, pig,

TABLE I
Comparison of Eukaryotic DHFR Sequences[a]

```
Fungi
PC .....MNQQKSLTLIVALTTSYGIGRSNSLPW.KLKKEISYFKRVTSFVPTFDSFESMNVVLMGRKTWESI

Yeast
SC ...MAGGKIPIVGIVACLQPEMGIGFRGGLPW.RLPSEMKYFRQVTSLTKDPNK...KNALIMGRKTWESI
CN MQTTAKSSTPSITAVVAATAENGIGLNGGLPW.RLPGEMKYFARVTTGETPSSDPSEQNVVIMGRKTWESI

Vertebrates
Ch ........VRSLNSIVAVCQNMGIGKDGNLPWPPLRNEYKYFQRMTSTSHVEGK...QNAVIMGKKTWFSI
Bo ........VRPLNCIVAVSQNMGIGKNGDLPWPPLRNEFQYFQRMTTVSSVEGK...QNLVIMGRKTWFSI
Mu ........VRPLNCIVAVSQNMGIGKNGDLPWPPLRNEFKYFQRMTTTSSVEGK...QNLVIMGRKTWFSI
Po ........VRPLNCIVAVSQNMGIGKNGDLPWPPLRNEYKYFQRMTTTSSVEGK...QNLVIMGRKTWFSI
Hu ........VGSLNCIVAVSQNMGIGKNGDLPWPPLRNEFRYFQRMTTTSSVEGK...QNLVIMGKKTWFSI
Ha ........VRPLNCIVAVSQNMGIGKNGDLPWPMLRNEFKYFQRMTTTSSVEGK...QNLVIMGRKTWFSI

          .         .         .         .         .         .
          10        20        30        40        50        60
```

```
Fungi
PC PLQFRPLKGRINVVITRNESLDLGNGIHSAK.S.LDHALELLYRTYGSESSVQINRIFVIGGAQLYKAAMD

Yeast
SC PPKFRPLPNRMNVIISRSFKDDFVHDKERSI..VQSNSLANAIMNLESNFKEHLERIYVIGGGEVYSQIF.
CN PSRFRPLKNRRRNVVISGKG.VDLGTAENSTVYTDIPSALSALRSTTESGHSP...RIFLIGGATLYTSSLL

Vertebrates
Ch PEKNRPLKDRINIVLSRELKEAPKGAHYLSK.S.LDDALALLDSPELKSKVD...MVWIVGGTAVYKAAME
Bo PEKNRPLKDRINIVLSRELKEPPKGAHFLAK.S.LDDALELIQDPELTNKVD...VVWIVGGSSVYKEAMN
Mu PEKNRPLKDRINIVLSRELKEPPRGAHFLAK.S.LDDALRLIEGPELASKVD...MVWIVGGSSVYEQAMN
Po PEKNRPLKDRINIVLSRELKEPPQGAHFLAK.S.LDDALKLTEGPELKDKVD...MVWIVGGSSVYKEAMN
Hu PEKNRPLKGRINLVLSRELKEPPQGAHFLSR.S.LDDALKLTEQPELANKVD...MVWIVGGSSVYKEAMN
Ha PEKNRPLKDRINIVLSRELKEPPQGAHFLAK.S.LDDALKLIEQPELADKVD...MVWIVGGSSVYKEAMN

          .         .         .         .         .         .
          70        80        90        100       110       120
```

```
Fungi
PC H....................PKLDRIMATIIYKDIHCDV.FFPLKFRDKEWSSVWKKEKHSD.......

Yeast
SC ....................SITDHWLITKINPLDKNAT.PAMDTFLDAKKLEEVFSEQDPAQLKEFLP
CN PSSVPSLNSSTSTSPLPFSFR.PLIDRILLTRILSPFECDAYLEDFAAHTKPDGSKVWKKASIKE......

Vertebrates
Ch K....................PINHRLFVTRILHEFESDT.FFPEIDYKDFKL................
Bo K....................PGHVRLFVTRIMQEFESDA.FFPEIDFEKYKL................
Mu E....................PGHLRLFVTRIMQEFESDT.FFPEIDLGKYKL................
Po K....................PGHIRLFVTRIMKEFESDT.FFPEIDLEKYKL................
Hu H....................PGHLKLFVTRIMQDFESDT.FFPEIDLEKYKL................
Ha Q....................PGHLRLFVTRIMQEFETDT.FFPEIDLEKYKL................

          .         .         .
          130       140       150
```

TABLE I (continued)

```
    Fungi
PC  ..LESWVGTKVPHGKINEDGFDYEFEMWTRDL

    Yeast
SC  KVELPETDC.DQRYSLEEKG..YCFEFTLYNRK
CN  ..FREWIGWDIEEQ.VEEKGVKYIFEMWVLNQ

    Vertebrates
Ch  ..LTEYPG..VPADIQEEDGIQYKFEVYQKSVLAQ
Bo  ..LPEYPG..VPLDVQEEKGIKYKFEVYEKNN
Mu  ..LPEYPG..VLSEVQEEDGIKYKFEVYEKKD
Po  ..LSECSG..VPSDVQEEKGIKYKFEVYEKNN
Hu  ..LPEYPG..VLSDVQEEKGIKYKFEVYEKND
Ha  ..LPEYPR..VLPEVQEEKGIKYKFEVYEKKG
        .           .       .
       160         170     180
```

[a] Abbreviations: PC = *P. carinii* (1); SC = *S. cerevisiae* (2,3); CN = *C. neoformans* (4); Ch = chicken (5); Bo = bovine (6); Mu = murine (7); Po = porcine (8); Hu = human (9); Ha = Chinese hamster (10). Residues identical for all nine species are shown in boldface.

and mouse sequences have relatively few differences, and many are highly conservative changes. However, the bovine sequence differs significantly more, and the chicken sequence has by far the most differences. The number of residues in the vertebrate sequences identical with corresponding residues in sequences of bacterial DHFRs (11) is about the same as the number of residues identical in the vertebrate and yeast sequences.

TABLE II
Numbers of Non-identical Residues in Pairs of Vertebrate DHFRs[a]

	Chicken	Bovine	Murine	Porcine	Human	Hamster
Chicken	0	44(9)	43(11)	41(8)	47(12)	47(11)
Bovine	44(9)	0	22(3)	19(2)	25(6)	21(3)
Murine	43(11)	22(3)	0	18(4)	21(8)	13(2)
Porcine	41(8)	19(2)	20(4)	0	21(8)	17(4)
Human	47(12)	25(6)	21(8)	24(8)	0	20(8)
Hamster	47(11)	21(3)	13(2)	17(4)	20(8)	0

[a] The three extra residues in the chicken sequence are not included in the counts. The numbers in parentheses are the numbers of the following conservative replacements: Y/F, S/T, K/R, I/V, V/L, I/L, or D/E.

B. SECONDARY AND TERTIARY STRUCTURE

Crystal structures have been reported by Kraut et al. for the binary complex of NADPH with chicken DHFR (12), and for ternary complexes of the chicken enzyme with NADPH and inhibitors (12–14), as well as with NADP and biopterin (15). Stammers et al. (16) reported structures for the ternary complexes of the mouse enzyme with NADPH and trimethoprim, and with NADPH and methotrexate. Structures for binary complexes of the human enzyme with folate, trimethoprim, and methotrexate have been published by Oefner et al. (17), while Davies et al. (18) published structures for binary complexes with folate and 5-deazafolate. The structure of the ternary complex of human DHFR with NADPH and MTX-γ-tetrazole has been reported by Cody et al. (19).

The general structure of the three vertebrate enzymes is quite similar. Thus when the structures of the chicken DHFR.NADPH complex and the human DHFR.folate complex are compared by least-squares superposition, the root-mean-square (rms) deviation in α-carbon positions is 0.6 Å for 182 residues (18). The largest positional deviations can be attributed to intermolecular lattice interactions in one or the other structure, or to regions of considerable flexibility.

The core of the human DHFR molecule is an eight stranded β sheet, consisting of seven parallel strands, and a carboxyl-terminal antiparallel strand (Fig. 1). The seventh strand, βG, is not continuous but consists of two segments, called βG1 and βG2 by Davies et al. (18), interrupted by a seven residue loop containing a tight turn. Residues involved in the β strands are identified in Table III. Five α helices (Table III) are packed against the β sheet. One of these, αe′, immediately follows αe, but has an axis nearly perpendicular to that of the latter. There is also one left-handed polyproline-like helix (residues 21–26), eight tight turns, and several extended loops.

1. Substrate-Binding Site

In the folate binary complex of hDHFR, both N3 and the 2-amino group of the pterin moiety are hydrogen bonded to the side chain carboxyl of Glu[30] (Fig. 2). The 2-amino group is also hydrogen bonded to bound water molecule 402, which is in turn hydrogen bonded to the side chain O of Thr[136]. The O4 of the pterin is hydrogen

Figure 1. The α-carbon representation of human DHFR complexed with folate. Two molecules with slight differences in structure are present in the asymetric unit. The diagram is for Molecule 2. The β strands are labeled with upper case letters, and α helices with lower case. [Reproduced with permission from Davies, J. F., Delcamp, T. J., Prendergast, N. J., Ashford, V. A., Freisheim, J. H., and Kraut, J., *Biochemistry*, **29**, 9467–9479 (1990), Copyright © 1990 American Chemical Society.]

29

TABLE III
Residues Involved in Secondary Structure of Chicken DHFR and Human DHFR

	Helices			β Strands	
	Human	Chicken		Human	Chicken
Pro(II)	21–26	23–26	βA	4–10	4–10
αb	27–40	28–40	βB	47–53	48–54
αc	53–59	55–60	βC	71–76	71–76
αe	92–102	92–101	βD	87–90	87–90
αe′	102–109		βE	108–116	111–116
αf	117–127	118–127	βF	130–139	131–139
			βG1	157–159	157–172
			βG2	168–172	
			βH	175–185	176–183

bonded to bound water molecule 401, which in turn is hydrogen bonded to OE2 of Glu[30], and weakly bonded to the indole N of Trp[24]. No specific interaction occurs between the enzyme and N1, N5, N8, or N10 of folate, but it is postulated that a hydrogen bond is likely between N8 of dihydrofolate and the carbonyl O of Ile[7]. Apart from these hydrogen-bond interactions of the pterin moiety with polar groups, there are nonspecific interactions with a number of hydrophobic side chains, including those of Leu[22], Phe[31], and Phe[34].

The p-aminobenzoyl glutamate (pABA) moiety of folate interacts nonspecifically with the side chains of five hydrophobic residues: Phe[31], Phe[34], Ile[60], Pro[61], and Leu[67]. There is also a hydrogen bond between the carbonyl O of pABA and the side chain amide N of Asn[64].

The α carboxylate of the glutamate moiety of folate interacts strongly with Arg[70], forming two hydrogen bonds as well as a salt bridge. On the other hand the γ carboxylate does not have strong interactions with any residue. No other residues make specific interactions with the glutamate moiety but the β-carbon atoms of Phe[31] and Arg[32], CE2 of Phe[34], and CD2 of Leu[67] make close contacts with carboxyl oxygen atoms of the glutamate moiety.

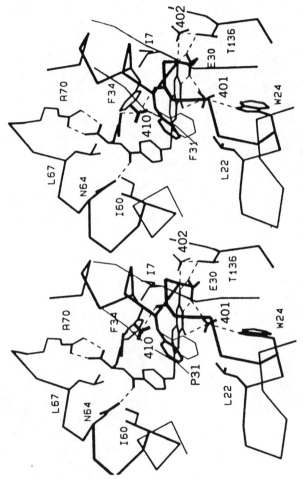

Figure 2. Interactions between human DHFR and folate (Molecule 2, see legend to Fig. 1), with hydrogen bonds depicted as broken lines. Water molecules are depicted as tetrahedra. For comparison, the orientation of Phe[31] in Molecule 1 is indicated by thin lines. [Reproduced with permission from Davies, J. F., Delcamp, T. J., Prendergast, N. J., Ashford, V. A., Freisheim, J. H., and Kraut, J., *Biochemistry*, **29**, 9467–9479 (1990). Copyright © 1990 American Chemical Society.]

31

2. Binding of Biopterin in the Substrate Site

McTigue et al. (15) reported the crystal structure of the ternary complex of chicken DHFR with NADP and biopterin (Fig. 3). The 7,8-dihydro form of biopterin is a substrate of human and other vertebrate DHFRs but with a much higher K_m and a lower k_{cat} than folate (20). The reduction product, tetrahydrobiopterin, is an important cofactor in the synthesis of biogenic amines. Biopterin lacks the pABA side chain, having instead a dihydroxypropyl side chain attached at the 6 position of the pteridine ring (Fig. 4). Despite this structural difference, this alternate substrate binds with the pteridine ring in virtually the same position in the active site as the pteridine ring of folate in its complex with human DHFR. The slight difference in the geometry, which will be discussed later, is attributed not to the difference in substrate side chains, but to the absence of dinucleotide in the complex of the human DHFR.

A unique feature of the chicken DHFR.NADP.biopterin complex is the presence of a bound water (Wat[756]) that is hydrogen bonded with O4, N5, and the two side chain hydroxyl groups of biopterin.

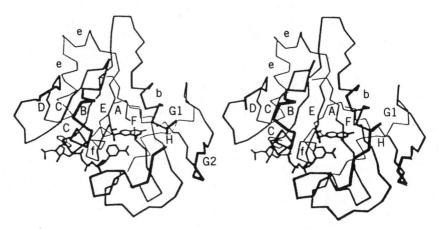

Figure 3. The α-carbon representation of chicken DHFR complexed with NADP[+] and biopterin (a folate analog). The β strands are labeled with upper case letters and α helices with lower case. [Reproduced with permission from McTigue, M. A., Davies, J. F., II, Kaufman, B. T., and Kraut, J., *Biochemistry,* **31,** 7264–7273 (1992). Copyright © 1992 American Chemical Society.]

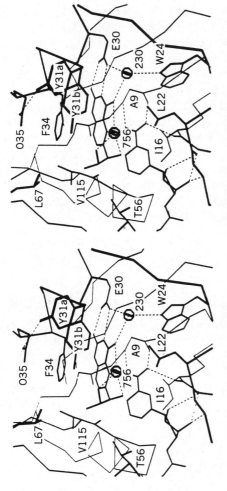

Figure 4. Catalytic site of chicken DHFR complexed with NADP⁺ and biopterin. Hydrogen bonds are depicted as broken lines, and water molecules are displayed as spheres. For clarity, only side chains that directly contact NADP⁺ or biopterin are shown. [Reproduced with permission from McTigue, M. A., Davies, J. F., II, Kaufman, B. T., and Kraut, J., *Biochemistry*, **31**, 7264–7273 (1992). Copyright © 1992 American Chemical Society.]

3. Binding of Methotrexate

Stammers et al. (16) reported the crystal structure of mouse DHFR.NADPH.MTX at 2.5 Å, and Oefner et al. (17) described the structure of the binary human DHFR.MTX complex at 3.5 Å. A structure for the ternary complex of human DHFR with MTX-γ-tetrazole and NADPH has been determined at 2.3 Å by Cody et al. (19), and since the γ-tetrazole ring is not in contact with the enzyme this is essentially equivalent to the ternary complex of MTX. In all three cases, the electron density is consistent with binding of the 2,4-diaminopteridine ring in the active site like its binding in the active site of bacterial DHFR (Fig. 5). A major difference between MTX binding and folate binding to human DHFR is that in the former the pteridine ring is "flipped over", that is, rotated 180° about the C6—C9 bond compared with the pteridine ring of folate. As a consequence of this the Glu^{30} carboxyl group hydrogen bonds with N1 and the 2-amino group of MTX, rather than with N3 and the 2-amino as in the case of folate. As in MTX complexes with bacterial DHFR, MTX bound to vertebrate DHFR is protonated as indicated by ultra-

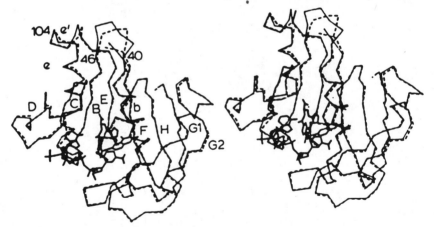

Figure 5.   Stereo comparison of the C$^\alpha$ trace of the human DHFR NADPH MTX-γ-tetrazole complex (solid line) with that of the folate complex of human DHFR (broken line). The backbone atoms of Glu^{30}, Phe^{34}, Thr^{136}, and Arg^{70} were used to align the structures. Also shown are NADPH and MTX–γ-tetrazole. [Reproduced with permission from Cody, V., Luft, J. R., Ciszak, E., Kalman, T. I., and Freisheim, J. H., *Anti-Cancer Drug Design, 7*, 483–491 (1992). Copyright © 1992 Oxford University Press.]

violet (UV) spectra (21, 22), calorimetric data (21), and nuclear magnetic resonance (NMR) results (23). Since the Glu[30] carboxyl group is presumably in its anionic form, the interaction with N1 of MTX is ionic in nature. There are also hydrogen bonds between the 2-amino group and Thr[136] via a bound water, and between the 4-amino group and the carbonyl oxygen atoms of Ile[7] and Val[115]. A structurally invariant water molecule forms a hydrogen-bond bridge from the carboxylate group of Glu[30] to the side chain of Trp[24] (as in the folate complex). In the ternary complex of the human enzyme, the benzoyl carbonyl forms hydrogen bonds with Asn[64] and with a fixed water molecule.

In the MTX complex of human DHFR the location of the pABG side chain is similar to that in MTX complexes with bacterial DHFRs and to that in the folate complex of human DHFR. There are similar interactions of hydrophobic residues with the benzoyl moiety, and the α-carboxylate forms two charge-mediated strong hydrogen bonds with the guanidinium ion of the conserved Arg[70].

The largest structural difference between human DHFR in the folate binary complex and in the MTX–tetrazole ternary complex is in the position of the flexible loop involving residues Thr[40] to Lys[46]. To determine whether this movement of the loop has significance for the binding of ligands at the active site will require structural data from additional complexes.

A surprisingly different orientation of the pABG side chain was observed in the mouse DHFR ternary complex (16). The difference involves substantial alterations in the torsion angles around some single bonds (C6—C9, C9—N10, N10—C4′, and φ of the peptide unit). As a result, the ionic interaction of the α-carboxylate with Arg[70] is lost, as are hydrophobic interactions with such side chains as Phe[34] and Leu[67]. Oefner et al. (17) view the mouse and human DHFR complexes with MTX as locked in different possible conformational states, but it is uncertain whether this arises from different conditions of crystallization, different crystal packing forces, or differences in the primary structure.

4. Binding of Phenyltriazine Inhibitors

The binding of six of these inhibitors to chicken DHFR has been investigated by Volz et al. (13), and by Matthews et al. (14). These inhibitors are all 2,4-diamino-5,6-dihydro-6,6-dimethyl-5-phenyl-*s*-triazines with a substituent at the 3 or 4 position of the phenyl ring. These substituents range in size and complexity from single atoms to large branching side chains. For all these inhibitors the binding of the triazine and phenyl rings is identical (Fig. 6). As in the case of bound MTX, N1 of the bound inhibitor is protonated with partial delocalization of positive charge onto the 2-amino group. There are hydrogen bonds and ionic interactions between these two protonated nitrogen atoms and the carboxylate group of Glu[30]. The 2,4-diamino-triazine moiety makes other hydrogen bond interactions and hydro-

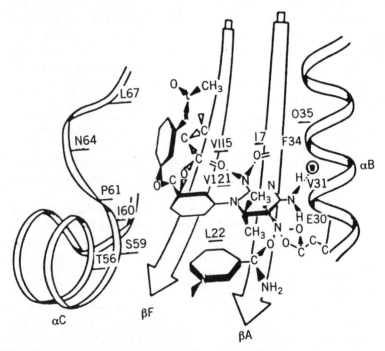

Figure 6. Schematic representation of phenyltriazine binding to chicken DHFR. Long chain 3′ substituents for two of the inhibitors are indicated by *blackened* and *open bonds,* respectively. Only the nicotinamide portion of NADPH is shown. A fixed water molecule is indicated by a large circle. Hydrogen bonds are represented by *dotted lines.* [Reproduced with permission from Matthews, D. A., Bolin, J. T., Burridge, J. M., Filman, D. J., Volz, K. W., and Kraut, J., *J. Biol. Chem.,* **260,** 392–399 (1985). Copyright © 1985 American Society for Biochemistry and Molecular Biology, Inc.]

phobic interactions with active site residues very similar to those made by the 2,4-diaminopteridine ring of MTX. Hydrophobic interactions are made by the phenyl ring with some of the same side chains that interact with the benzoyl moiety of MTX (Leu[22], Phe[31], Phe[34], and Ile[60]).

Inhibitors with large substituents at the 3 position of the phenyl ring accommodate these long side chains in the narrow channel leading from the active site cleft to the protein surface, that is bounded primarily by Tyr[31], Phe[34], Ile[60], Pro[61], and Leu[67]. This is the same channel in which the pABG side chain of folate is accommodated.

In the complex of 2,4-diamino-5-(1-adamantyl)-6-methylpyrimidine, which binds quite tightly to chicken DHFR, the bulky adamantyl moiety is accommodated in the same position as the phenyl group of the phenyltriazines, and makes good van der Waals contacts with the surrounding hydrophobic side chains.

5. Binding of Trimethoprim

The structure of the complex of this inhibitor with vertebrate DHFR is of particular interest because the binding is so much weaker than trimethoprim binding to bacterial DHFR, 30,000 times weaker for human DHFR than for *Escherichia. coli* DHFR (24). When the trimethoprim complexes of chicken DHFR and *Escherichia. coli* DHFR are compared the following differences are apparent (12, 14). In the chicken DHFR complex the trimethoxybenzyl side chain of trimethoprim binds in the same position as the phenyl or adamantyl groups of the better binding inhibitors, and is directed "up", that is, away from the nicotinamide binding site (Fig. 7). The pyrimidine ring is in approximately the same position as for the adamntyl derivative, but slightly tilted in relation to that of the latter. However, accommodation of the methylene group of the trimethoprim side chain forces the pyrimidine ring approximately 1 Å closer to helix αb, with the result that no hydrogen bond is possible between the 4-amino group and the carbonyl oxygen of Val[115]. Such a bond is present in the complexes of all tightly bound inhibitors. In *E. coli* DHFR, the active site cleft is 1.5–2.0 Å narrower than for vertebrate DHFR, so that the trimethoxybenzyl side chain cannot locate in the "upper" channel leading from the active site to the surface of the protein, but instead is directed down towards the nicotinamide-binding site where the cavity is wider (Fig. 8). In this position the benzyl

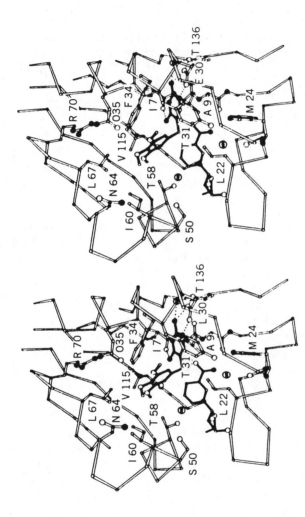

Figure 7. The binding of trimethoprim and NADPH to chicken DHFR. Trimethoprim and a portion of the NADPH molecule are represented by *solid bonds*, and protein by *open bonds*. Carbon atoms are represented by smaller *open circles*, oxygen atoms by *larger open circles*, and nitrogen atoms by *blackened circles*. *Large numbered circles* represent fixed water molecules. Hydrogen bonds, are indicated by *broken lines*. [Reproduced with permission from Matthews, D. A., Bolin, J. T., Burridge, J. M., Filman, D. J., Volz, K. W., Kaufman, B. T., Beddell, C. R., Champness, J. N., Stammers, D. K., and Kraut, J., *J. Biol. Chem.*, **260**, 381–391 (1985). Copyright © 1985 American Society for Biochemistry and Molecular Biology, Inc.]

38

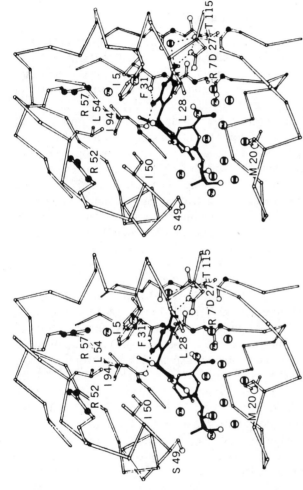

Figure 8. The binding of trimethoprim to *E. coli* DHFR with NADPH modeled into the active site. Trimethoprim and a portion of the NADPH molecule are represented by *solid bonds*, and protein by *open bonds*. Carbon atoms are represented by *smaller open circles*, oxygen atoms by *larger open circles*, and nitrogen atoms by *blackened circles*. *Large numbered circles* represent fixed water molecules. Hydrogen bonds are indicated by *dashed lines*. [Reproduced with permission from Matthews, D. A., Bolin, J. T., Burridge, J. M., Filman, D. J., Volz, K. W., Kaufman, B. T., Beddell, C. R., Champness, J. N., Stammers, D. K., and Kraut, J., *J. Biol. Chem.*, **260**, 381–391 (1985).]

group makes favorable hydrophobic interactions with Ile[50] and Leu[28]. Because of the wider active site cavity in chicken DHFR, interactions with Ile[60] and Tyr[31], the corresponding residues, would be far less favorable with the benzyl side chain in this conformation.

6. Nucleotide-Binding Site

In the ternary complex of NADPH and MTX-γ-tetrazole with human DHFR, NADPH is bound in an extended conformation with the 2'-phosphoADP-ribose moiety occupying a long shallow cleft that covers the carboxy terminal ends of five strands of β sheet (Fig. 5). The nicotinamide moiety is inserted between the carboxy ends of βA and βE, while both ribosyl groups are more exposed to solvent. The carboxamide group of the nicotinamide moiety is syn to the nicotinamide ring N, and makes a series of strong hydrogen bonds to the conserved residues Ala[9] and Ile[16]. There are also several non-bonded C—H⋯O contacts involving the nicotinamide ring carbon-atoms and three neighboring oxygen atoms of Ile[16], Thr[56], and Val[115], which lie approximately in the plane of the nicotinamide ring. The adenine ring occupies a nonspecific cleft between the side chains of Leu[75] and Arg[77] on the opposite side of the β sheet to the active site cleft.

In all these respects the interaction of NADPH with human DHFR is very similar to the binding of NADPH and NADP to chicken DHFR (13, 15) and to bacterial DHFRs. In the NADP and NADPH complexes of the chicken enzyme nearly all the residues that interact with the bound dinucleotide are conserved, and are identical for NADP and NADPH. The adenine residue makes hydrophobic contacts with the side chains of Leu[75], Ala[119], and Val[120], and with the backbone of Ser[92]. The ADP ribose makes hydrophobic interactions with the backbones of Lys[54], Leu[75], and Ser[76]. The 2'-phosphate makes charge interactions with the amino of Lys[54] and (via bound Wat[393]) with the guanidinium of Arg[77]. In addition, there are hydrogen bonds from the Arg[77] backbone amido to the 2'-phosphate and from the Lys[55] amido to one of the pyrophosphate oxygen atoms.

The pyrophosphate group binds at the carboxyl end of the β sheet near the amino termini of αc and αf, where its negative charge is stabilized by the positive ends of the two helix dipoles (Fig. 3). The pyrophosphate moiety leads over the end of the β sheet to the nicotinamide binding site on the "front" side of the β sheet. The pyrophosphate group is bound by hydrogen bonds to Thr[56] (amido and hydroxyl), Gly[117] (amido), Ser[118] (hydroxyl and amido), and Thr[146] (hydroxyl). There are also hydrophobic interactions with the Lys[55] side chain and with Gly[117], and a charge interaction with the Lys[55] amino group.

The nicotinamide ring is bound near the center of the enzyme structure in a cleft formed by the divergence of βE and βA. A conserved cis-peptide bond between Gly[116] and Gly[117], which terminates βE creates a curve in the backbone that accommodates the nicotinamide ring between this cis-peptide bond and Ala[9] at the C-terminus of βA, in a plane parallel with the β sheet. The B side of the nicotinamide ring faces the interior of the protein, and is packed against the side chains of Ile[16] and Tyr[121]. In the binary complex, the A side is exposed to solvent, but in ternary complexes it is close to the pyrazine part of the substrate or inhibitor pteridine ring. The carboxamide group is kept approximately in the plane of the nicotinamide ring by three hydrogen bonds. The carboxamide NH_2 hydrogens bond to backbone oxygen atoms of Ala[9] and Ile[16], and its carbonyl hydrogen bonds to the backbone N of Ala[9] (Fig. 4). There are hydrophobic interactions between the nicotinamide ring and the side chains of Ile[16], Leu[22], and Trp[24], and the backbone of Val[8].

The nicotinamide ribose makes hydrophobic interactions with Gly[20], Gly[117], and the side chains of Ile[16], and Thr[56]. There are also hydrogen bonds between the backbone carbonyl and amido groups of Asn[21] and the 2'- and 3'-oxygen atoms of this ribose.

Residues of chicken DHFR that are involved in these interactions are invariant in vertebrate DHFRs with the following exceptions: Ala[119] is a Ser in the others; Lys[54] is Arg in others except human; Thr[146] is Ser in bovine; Asn[21] is Asp in others.

7. Comparison of the NADPH-Binding Site of DHFR with the NAD-Binding Site of Dehydrogenases

The NAD-binding site of lactate dehydrogenase shares a number of features with the NADPH-binding site of DHFR (25): binding of the dinucleotide at the carboxyl end of a β-pleated sheet near the innermost strands; binding of the adenosine moiety on the back side of the β sheet; winding of the pyrophosphate moiety over the end of the β sheet; and stabilization of the negative charge on the pyro-phosphate group by the proximity of the positive end of the αb helix in lactate dehydrogenase and of the αc and αf helices in DHFR. There is also a structural analogy between the conservation of Gly[116] of vertebrate DHFR (and the corresponding Gly of bacterial DHFRs), which is in van der Waals contact with the pyrophosphate group of bound NADPH, and the conservation of a geometrically equivalent Gly of lactate dehydrogenase to avoid too close contact between this residue and the adenosine ribose of NAD. Similarly, the conservation of an Asp in lactate dehydrogenase to make a hydrogen bond with the 2'-hydroxyl of the adenosine moiety of NAD, is paral-lel to the conservation of Lys[54] (or an Arg) in vertebrate DHFR to form a charge interaction with the 2'-phosphate of bound NADPH.

The most obvious difference between the DHFR dinucleotide-binding site and that of the dehydrogenases is in the specific pattern of folding of strands of the β sheet and helices, as shown in Figure 9. The pattern for the dehydrogenases is conserved among many enzymes binding nucleotides (26).

8. Comparison of Structures of Vertebrate and Bacterial DHFRs

The secondary and tertiary structures of the vertebrate enzymes are quite similar to that of the bacterial DHFRs, despite the fact that the sequence identity is less than 30%. When the sequences are aligned according to elements of secondary structure as determined by crystallography, the following conclusions can be drawn (13). About 70% of the additional residues in the vertebrate enzymes occur in three loops far removed from the active site. The two inser-tions with major affects on secondary structure are as follows. The Pro[25], inserted in the loop joining βA and αb, results in a segment Pro[23]-Trp[24]-Pro[25]-Pro[26] that forms an almost ideal polyproline three-fold helix. This structure, present in all except one of the known vertebrate sequences, is absent from bacterial DHFRs, where the chain is almost fully extended at this point. The significance of this structure is unclear. The second such insertion is that of six residues in the middle of the βG strand, where as noted above, they contribute to a seven residue loop or bulge in the βG strand.

Despite the high degree of overall structural homology between the vertebrate and bacterial enzymes, there are differences of 1–3

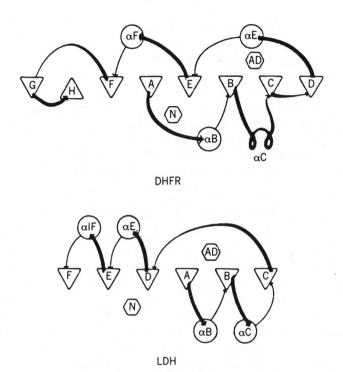

DHFR

LDH

Figure 9. *Upper*, schematic diagram showing how individual strands are connected to form the central β-pleated sheet in DHFR and how NADPH binds in relation to this sheet. Each strand is represented as either ▽ or △ depending on whether the carboxyl or amino end is nearer the viewer. Helices with their axes nearly parallel to the adjacent β sheet are indicated by *circles*, while the one helix (αC), which is almost perpendicular to adjacent β sheet is depicted as a two-turn helix. Connections occurring at the end of the sheet nearest the viewer are drawn with *thick lines*. Hexagons labeled N and AD represent the nicotinamide and adenine mononucleotide portions of NADPH, respectively. *Lower*, schematic diagram for the NAD$^+$-binding region in lactate dehydrogenase (LDH). [Reproduced from Matthews, D. A., Alden, R. A., Bolin, J. T., Filman, D. J., Freer, S. T., Hamlin, R., Hol, W. G. J., Kisliuk, R. L., Pastore, E. J., Plante, L. T., Xuong, N.-H., and Kraut, J., *J. Biol. Chem.*, **253**, 6946–6946 (1978), with permission of the authors.]

Å in the relative placement of corresponding elements of secondary structure (13, 14). In particular residues on opposite sides of the active site cleft of the vertebrate enzyme are 1.5–2.0 Å further apart than are structurely equivalent residues in the bacterial enzyme. This separation is probably the direct result of a three-residue insertion in the vertebrate sequence immediately following the termination of helix αc, so that in the vertebrate sequence the six residues following this insertion are displaced slightly further away from helix αb on the opposite side of the cleft. This slight expansion of the cleft in the vertebrate enzyme is greatest between Glu[30] and Ile[60] (2.05 Å in chicken DHFR), and the longer side chain of the invariant Glu[30] in vertebrate DHFR, instead of Asp in the corresponding position in bacterial DHFR, compensates for the expanded cleft. The expanded cleft is also directly related to the greatly decreased affinity of vertebrate DHFR for certain inhibitors like trimethoprim.

9.   Structural Changes Induced by NADPH Binding

One structural effect of dinucleotide binding to vertebrate DHFR can be discerned by comparison of the crystal structure of chicken DHFR apoenzyme with that of the chicken DHFR.NADPH binary complex, termed "holoenzyme" by the crystallographers (27). The two sets of crystals are not isomorphous, suggesting of itself that conformational changes occur on binding NADPH. The most striking structural change is a rearrangement of the polypeptide backbone and side chain atoms in the neighborhood of the adenine binding site, particularly for residues 91–96. In the apoenzyme these residues form an irregular loop connecting βD to αe, which is a five-residue helix. As a result of NADPH binding these residues move over to cap the adenine-binding pocket, making extensive van der Waals contacts along the exterior edge of the pyrimidine portion of the ring. Moreover in this complex, five additional residues (residues 91–96) become a part of αe. Backbone movements for residues 91 and 92 are about 6 Å. There are smaller back bone adjustments (1–2 Å) for residues 76 and 77.

Even larger rearrangements of some of the side chains occur with NADPH binding. Thus the side chain of Lys[91] is displaced 15 Å, and the Arg[77] side chain is displaced 3 Å to form one side of the adenine binding cleft. The side chain of Lys[54] is disordered in the apoenzyme, but in the dinucleotide complex it is hydrogen bonded to the 2′-phosphate. Also, as positioned in the apoenzyme, the side chain of Lys[55] would sterically hinder NADPH binding. In the dinucleotide complex this side chain is repositioned (5.8 Å movement for Nζ) to form a charge-mediated hydrogen bond with the pyrophosphate moiety.

In contrast to these changes in the vicinity of the adenine and pyrophosphate binding regions, there are no clear indications of conformational changes in the nicotinamide-binding site or adjacent regions of the active site of chicken DHFR when NADPH binds.

When the crystal structure of the binary folate complex of human DHFR is compared with the ternary complex with NADPH and MTX-γ-tetrazole (Fig. 5), some structural changes are also observable, but they are not as extensive as those described for the comparison of the chicken apoenzyme and its NADPH complex. The only region of backbone that is significantly moved in the ternary structure of the human enzyme is the flexible loop at residues 40–46, but the significance of this is uncertain. As in the case of chicken DHFR, Lys[54] is hydrogen bonded to the 2'-phosphate of NADPH, but Lys[55] is not hydrogen bonded to any of the phosphates, and there is much less movement of Arg[91] when NADPH binds to human DHFR than for Lys[91] of chicken DHFR. Moreover, the only atom of residues 91–96 of human DHFR that makes van der Waals contact with the adenine ring of bound NADPH is the oxygen of Arg[91]. Whether these different effects of NADPH binding reflect the influence of different crystal packing forces, or the presence of ligand in the substrate site in the case of human DHFR but not in that of chicken DHFR, or to species differences, is not clear.

10. Structural Changes Induced by Substrate Binding

The effect of substrate binding may be inferred by comparing the structures of the chicken DHFR.NADPH complex with the chicken DHFR.NADP.biopterin complex (15). Binding of biopterin to the dinucleotide complex appears to have only a negligible effect on the position of the protein backbone. However, there is a decrease in mobility of the protein structure, as indicated by a decrease in the average thermal factor B from 35 Å^2 for the dinucleotide complex to 22 Å^2 in the biopterin ternary complex.

In the ternary structure, the position of the side chain carboxylate of Glu[30] is stabilized by hydrogen bonds to N3 and the 2-amino group of biopterin (B decreased from 74 to 9 Å^2), and there is an associated increase in the ordering of Wat^{230}, which is hydrogen bonded to the carboxylate of Glu^{30} and to the indole N of Trp^{24}.

The other major difference between the binary and ternary structures is that the side chain of Tyr^{31} occupies two alternative positions in the ternary structure (Fig. 4). In one conformation (30% occupancy), which is similar to that in the dinucleotide binary complex, its side chain is hydrogen bonded to the side chain amide of Gln^{35}, and in the second conformation (70% occupancy) this side chain forms parts of the pteridine and Wat^{756} binding sites. It should be noted that Phe^{31} is present in different conformations in the two molecules in the assymetric cell for the human DHFR.folate complex (Fig. 2), so that this mobility of the side chain of residue 31 is not confined to ternary complexes of vertebrate DHFR.

In the ternary complex the nicotinamide ring of NADP is rotated slightly farther away from the substrate-binding site, so that the separation between C4 (nicotinamide) and C6 and C7 (pteridine) becomes 3.5 and 3.3 Å, respectively. The rotation is 9° about an axis collinear with the nicotinamide C2—C3 bond.

11. Structural Changes Due to Derivatization with Organomercurials

Vertebrate DHFRs can be activated by derivatization of the single Cys residue with organomercurials, the largest activations being 8- and 12-fold for derivitization of chicken DHFR with mercuribenzoate and methyl mercury, respectively (27a). Examination of the crystal structures of complexes of these derivatives by McTigue (27b) showed only small differences from the structures of the corresponding complexes of native enzyme. These small structural changes are localized to the modification site and are due to the insertion of the organomercurial. Activation is thought to be due to an increased rate of tetrahydrofolate release, which in turn is probably due to the global increase in thermal factors. Thus the average value of B for the protein in dinucleotide binary complexes is 35.2, 40.6, and 42.5 $Å^2$ for native enzyme, methyl mercury derivative, and mercuribenzoate derivative, respectively. The increase in B is even greater for the DHFR.NADP.biopterin complexes, 13.4 $Å^2$ for the methyl mercury modified, and 10.0 $Å^2$ for the mercuribenzoate. B factors for the dinucleotide and biopterin are increased about the same amount as those for the protein. The mechanism by which the motion of the entire protein and bound ligands is increased is uncertain, but may be related to an effect on βA (residues 4–10) one of the central strands in the β sheet. The residue substituted is either adjacent to this strand (Cys[11] in chicken DHFR) or in it (Cys[6] in mammalian DHFRs). Effects on the position and motion of Trp[24] may also be involved.

C. EVIDENCE OF STRUCTURE FROM NMR

Stockman et al. (28) used heteronuclear three-dimensional (3-D) NMR spectroscopy to make sequential resonance assignments for more than 90% of the residues in human DHFR complexed with MTX. Uniform enrichment of the protein with ^{15}N was required to obtain the resonance assignments via heteronuclear 3-D NMR spectroscopy since homonuclear 2-D spectra did not provide sufficient ^1H resonance dispersion. Medium- and long-range nuclear overhauser effects (NOEs) are consistent with the existence in solution of the eight-stranded β sheet evident in the crystal structure. However, no further structural details that this method can provide have yet been published. There appears to be no evidence for two interconverting isomers like that detected by NMR spectroscopy on the MTX complex of E. coli DHFR (29).

III. Kinetics and Thermodynamics of Ligand Binding

Since the rates of substrate binding and of product release are the steps that limit the rate of turnover in the catalytic cycle for DHFR, at least in the physiological pH range, the corresponding rate constants are important for quantitative description of enzyme function. Equilibrium dissociation constants for these processes are also of significance in any complete understanding of the enzyme. Kinetic and thermodynamic parameters for inhibitor binding are similarly essential for comparison of inhibitors, and are a prerequisite for the description of inhibitor binding in molecular terms.

Dissociation constants (K_d) are usually determined fluorimetrically, by measuring the quenching of enzyme fluorescence when known concentrations of ligand are present. This quenching of fluorescence also provides the basis for the measurement of association rate constants (k_{on}) with a stopped flow spectrofluorimeter. Such binding kinetics data can also give the dissociation rate constant (k_{off}), provided the value lies in a suitable range. However, k_{off} is usually better determined by the competition method, in which the dissociating ligand (L_1) is replaced by another ligand (L_2). For this method to give reliable results, L_1 must be present at a concentration sufficient to complex most of the enzyme, but low enough that L_2, added at a relatively high concentration, can compete successfully with L_1 for free enzyme. There must be a suitable method for measuring the rate of the decline in the concentration of E.L_1 or the rate of increase of E.L_2. The relationship between the true k_{off} and k_{obs}, the rate constant for the observed change, under various conditions of concentrations and rate constants has been described in detail (20, 30).

A. BINDING OF PHYSIOLOGICAL LIGANDS

1. Binding of NADPH and NADP

The binding of the dinucleotides NADPH and NADP have been studied in detail for only three eukaryotic DHFRs: human, mouse and that of the pathogenic fungus *Pneumocystis carinii*. However, thermodynamic studies have been reported for binding to chicken DHFR. Reported values for the thermodynamic dissociation constants, all determined by fluorescence quenching, are given in Table IV. In all cases NADP binds less tightly than NADPH. In the case of NADPH binding there is a 37-fold range in the values, which is somewhat surprising in view of the great similarity of the structures of their binding pockets. However, the K_d for NADPH binding to mouse DHFR was determined in the presence of 0.8 M NaCl (32), whereas all the other values were obtained at lower ionic strength. We have shown that the addition of 0.1 M NaCl, 20 mM 2-morpholinoethanesulfonic acid (MES) and 25 mM acetate to Tris buffer at pH 7.4 increases K_d for NADPH binding to human DHFR by a factor of 5 (20, cf. 33), so that further increasing the NaCl concentration to 0.8 M probably accounts for the high K_d obtained for mouse DHFR.

Reported values for k_{on} and k_{off} are shown in Table V. Although comparison of values of rate constants for formation of similar complexes of the human and mouse enzymes is possible in only a few cases, the values of k_{off} differ by only relatively small factors (up to

TABLE IV
Thermodynamic Dissociation Constants for Binary Complexes of Dinucleotides

Ligand	Species	K_d (nM)	Reference
NADPH	Human	50 ± 10[a]	31
	Mouse	$1,850 \pm 160$[b]	32
	Chicken	310[c]	21
	P. carinii	90 ± 9[d]	24
NADP	Human	$2,300 \pm 500$[a]	31
	Mouse	$3,710 \pm 220$[b]	32
	Chicken	$3,800$[c]	21
	P. carinii	$24,000 \pm 3,000$[d]	24

[a] pH 7.65, 20°.

[b] pH 7, 25°.

[c] pH 7.4, 24°.

[d] pH 7.0, 20°.

TABLE V
Rate Constants for Association and Dissociation of Dinucleotides with DHFR[a]

Ligand	DHFR Source	Enzyme Form[b]	k_{on} ($\mu M^{-1}s^{-1}$)	k_{off}^{c} (s^{-1})	k_{off}/k_{on} (μM)	K_d (μM)
NADPH	Human	E	38 ± 3	1.7 ± 0.1	0.04	0.05
	Mouse	E	2.8 ± 0.1	2.8 ± 0.3	1.00	1.85
	P. carinii	E	46 ± 3	4.93 ± 0.02	0.11	0.09
	Human	E.folate	29 ± 1	8.4	0.29	
	Human	E.H$_2$folate	24 ± 1	(19)[d]	0.79	
	Human	E.H$_4$folate	4.4 ± 1.6	100 ± 5	22.7	
	Mouse	E.H$_4$folate	1.0 ± 0.4	26 ± 15	26	
	Mouse	E.MTX	2.7 ± 0.1	2.2 ± 0.6	0.85	
NADP	Human	E	17 ± 1	32 ± 1	1.9	2.3
	Mouse	E	6.8 ± 0.2	90 ± 10	13.2	3.7
	P. carinii	E	7 ± 2	323 ± 73	46	24
	Human	E.folate	5.2	0.050	0.0096	
	Human	E.H$_2$folate	20 ± 1	4.6 ± 0.4	0.23	
	Mouse	E.H$_2$folate	2 ± 0.15	7.3 ± 0.9	3.7	
	P. carinii	E.H$_2$folate	10.3 ± 0.6	156 ± 9	15	
	Human	E.H$_4$folate	0.7 ± 0.1	84 ± 2	120	
	Mouse	E.H$_4$folate	9 ± 0.4	440 ± 40	48.9	
	P. carinii	E.H$_4$folate	>7	>1000		139

[a] References and experimental conditions as in Table IV.

[b] Form of the enzyme to which the dinucleotide binds.

[c] Values are those obtained by competition, where available.

[d] Obtained indirectly.

4), but values for k_{on} differ by factors of up to 13.6, and k_{off}/k_{on} values differ by factors of up to 25. The presence of 0.8 M NaCl for determinations of constants for mouse DHFR may have contributed to differences between some of these constants and corresponding constants for the other DHFRs.

2. Binding of Substrates and Tetrahydrofolate

The only eukaryotic DHFRs for which binding of pteridine substrates and products has been studied are again the human, mouse, chicken, and *P. carinii* enzymes. The K_d values for binding of the various pterins to human DHFR indicate quite strong binding (Table VI), with the exception of H_2biopterin, which binds very weakly. Reported binding of folates to mouse DHFR is significantly weaker than to the human enzyme (perhaps due to the high NaCl concentration present for the mouse DHFR studies), with the greatest decrease in the binding of folate itself, but chicken DHFR binds folate and H_2folate even more tightly than the human enzyme does.

TABLE VI

Thermodynamic Dissociation Constants for Binary Complexes of Pteridine Substrates and Tetrahydrofolate[a]

Ligand	DHFR Source	K_d (nM)	Reference
Folate	Human	83 ± 10	31
	Mouse	1,660 ± 70	32
	Chicken	67	21
	Chicken	830 ± 130[b]	34
	P. carinii	104 ± 14	24
H_2folate	Human	120 ± 40	31
	Human	150[c]	35
	Mouse	810 ± 70	32
	Chicken	14	21
H_2biopterin	Human	31,000 ± 5,000	20
H_4folate	Human	50 ± 14	31
	Mouse	200 ± 10	32

[a] Results of fluorescence titration, except where indicated. Experimental conditions as indicated for corresponding references in Table IV.

[b] pH 7.4; 30°.

[c] By equilibrium dialysis; pH independent.

Table VII gives the k_{on} and k_{off} values for some binary and ternary complexes of substrates and H_4folate. Since, wherever available, the k_{off} values in Table VII are from the competition method, k_{on}/k_{off} should agree with K_d. It may be seen that where the comparison is possible, the agreement is quite good (k_{off} obtained from binding kinetics does not include the effects of an isomerization of the complex prior to dissociation, should such an isomerization occur.) It may be seen that the very weak binding of H_2biopterin compared with H_2folate is due primarily to the very high rate of dissociation, and this in turn is doubtless due to the much smaller substituent on the pterin ring at position 6. This small side chain both decreases the number of interactions of this substrate with the protein, and also facilitates diffusion of this ligand out of the active site cavity.

TABLE VII

Rate Constants for Association and Dissociation of Substrates
and Tetrahydrofolate with DHFR[a]

Ligand	DHFR Source	Enzyme Complex[b]	k_{on} ($\mu M^{-1}s^{-1}$)	k_{off}[c] (s^{-1})	k_{off}/k_{on} (μM)	K_d (μM)
Folate	Human	E	156 ± 8	30	0.19	0.083
	Human	E.NADP	88	0.12	0.0014	
	Human	E.NADPH	58 ± 2	10	0.17	
H_2folate	Human	E	264 ± 8	14 ± 1	0.053	0.12
	Mouse	E	33 ± 5	20 ± 2	0.61	0.81
	P. carinii	E	27 ± 1	4 ± 0.1	0.18	0.10
	Human	E.NADP	110 ± 6	1.3 ± 0.1	0.012	
	Human	E.NADPH	98 ± 25			
H_2biopterin	Human	E	40 ± 16	2045 ± 532	51	31
H_4folate	Human	E	17 ± 6	5.1 ± 0.1	0.053	0.05
	Mouse	E	33 ± 2	15 ± 0.5	0.45	0.20
	P. carinii	E	25 ± 3	1.7 ± 0.1	0.065	
	Human	E.NADP	24 ± 2	46 ± 2	1.92	
	Mouse	E.NADP	18 ± 2	15 ± 0.5	0.93	
	P. carinii	E.NADP	37 ± 5	14 ± 0.9	0.38	
	Human	E.NADPH	14 ± 4	225 ± 25	16.1	
	Mouse	E.NADPH	5 ± 3	20 ± 1	4.0	
	P. carinii	E.NADPH		216 ± 4		

[a] References and experimental conditions as in Table VI.

[b] Complex to which the ligand binds.

[c] Values are those obtained by the competition method where available.

Values of k_{off} for various complexes vary by a factor of over 1500, and account for much of the differences in binding affinity, but there are also quite significant differences in k_{on}. The latter play a significant part, for example, in the weaker binding of H_2folate and H_4folate to mouse DHFR than to human DHFR. Again, however, this may be due to the high NaCl concentration present for determination of the mouse DHFR constants.

3. Effect of a Ligand Bound in One Site on Ligand Binding in the Other Site: Cooperativity of Ligand Binding

In Tables V and VII, the k_{off}/k_{on} values are estimates of K_d, and it may be readily seen that for a particular ligand and DHFR these values vary greatly, depending on whether NADP or NADPH is bound to the enzyme. Thus in the case of the human DHFR, binding of NADP to the enzyme with no other ligand bound (referred to by some authors as "apoenzyme") is about eight times weaker than binding to the DHFR.H_2folate complex, and about 63 times stronger than binding to the DHFR.H_4folate complex. Bound NADP similarly increases the binding of H_2folate and decreases the binding of H_4folate as measured by k_{off}/k_{on} values. The largest effect of this kind reported is that for the interaction of NADPH and H_4folate as ligands of human DHFR, where the mutual decrease in k_{off}/k_{on} values is by a factor of about 400.

A mutual increase in ligand affinity for the enzyme in the ternary complex has been called "synergism" by some authors, but this term is unsatisfactory because when the interaction of the bound ligands mutually *decreases* their binding energy "negative synergism" or "antagonism" are unsatisfactory terms for such behavior. A number of authors prefer the term cooperativity for this phenomenon (30, 36–39), and the cooperativity can be defined as k_{off}/k_{on} for the binary complex divided by k_{off}/k_{on} for the ternary complex. A cooperativity greater than 1 (positive cooperativity) indicates mutually enhanced binding, whereas cooperativity less than 1 (negative cooperativity) indicates mutually decreased binding. Examination of Tables V and VII shows that effects on k_{on}, k_{off}, or both may be responsible for cooperative effects of one ligand on the binding of another.

The molecular basis for this ligand-binding cooperativity is not clear. Bystroff and Kraut (39) considered the possible molecular basis for cooperativity of ligand binding to *E. coli* DHFR. Crystallographic evidence suggests that when a ligand binds to either the pterin site or the dinucleotide site of *E. coli* DHFR there is a conformational change that closes the pABGA-binding cleft and realigns α-helices c and f, which bind the pyrophosphate group of the dinucleotide. According to this view energy is expended in the binding of the first ligand to produce this conformational change, whereas in the binding of the second it is not. This binding gives rise to positive cooperativity. It is suggested that a second source of positive cooperativity could be completion of the binding site of one ligand by the other. Negative cooperativity, it is proposed, is due to overlapping of the binding sites, the extent of this crowding being directly proportional to the number of hydrogen atoms attached to the nicotinamide and pterin rings.

However, the crystallographic results for vertebrate DHFRs (Sections II.B.9 and 10) indicate that the conformational changes that occur on ligand binding to vertebrate DHFRs are so subtle that expenditure of significant energy to effect them seems unlikely. Thus it seems unlikely that the positive cooperativity that occurs for the binding of NADP and H_2folate to human DHFR arises from the small conformational changes revealed by crystallography. For human DHFR, there is also incomplete correlation between the number of hydrogen atoms attached to ligands and the type and extent of cooperativity. Thus NADPH and folate show strong negative cooperativity, whereas NADP and H_2folate display marked positive cooperativity. These apparent discrepancies between the hypothesis of Bystroff and Kraut and experimental results are probably due to conformational changes that are associated with the binding of specific ligands that have not yet been detected by crystallography or NMR.

4. Apoenzyme Conformers with Different Affinity for Physiological Ligands

As in the case of the bacterial DHFRs, vertebrate DHFRs exhibit kinetic behavior that points to the presence in solution of two conformers of the apoenzyme that are in relatively slow equilibrium (30, 31). The equilibrium constant (K_{eq}) for the conversion of the nonbinding form of human DHFR (E′) to the tight-binding form (E) is 2.4, whereas for the mouse enzyme it is 1. The apparent rate constant for the conversion of E′ to E is 0.09 and 0.048 s^{-1} for human and mouse DHFR, respectively. Whether NADPH can slowly bind to the "nonbinding" form cannot be determined from the kinetic data, which merely establish that the rate of such binding is slower than isomerization to the tight-binding form. Other ligands also bind more tightly to the E form of human DHFR than to the E′ form (31).

The rate of conversion of E′ to E is very much faster that the corresponding conformational change in bacterial DHFRs, which has a half-time of about 9 s (40, 41). Moreover, in the case of the bacterial DHFRs both NADPH and H_2folate bind exclusively to E.

In the case of mouse DHFR the data regarding this equilibrium between apoenzyme conformers is complicated by several factors. The first is that experiments were all performed at very high (0.8 M) sodium chloride concentration, which probably decreases the binding of the dinucleotides. (Compare the K_d, k_{on}, and k_{off} for human and mouse DHFRs in Tables IV and V.) Furthermore, the data are interpreted by the authors to mean that, under the experimental conditions, NADP binds *more tightly to E′ than to E*. In addition, although it can be calculated from the data (32) that the equilibrium constant for the conversion of E.NADP to E′.NADP is 24, it is necessary to assume that the binding of H_4folate to the NADP binary complex yields E.NADP.H_4folate, not E′.NADP.H_4folate.

Although consideration of crystallographic evidence (Section II.B.9) permits some speculation about the nature of the structural differences between the two forms, no firm conclusion is possible. It is perhaps significant, however, that the mutation causing the very conservative replacement of Trp[24] in human DHFR by Phe causes a marked change in the equilibrium between E and E′, with K_{eq} changing from 4 to 0.32 (42). This change suggests that the conformational change either involves residues in the Pro(II) helix, residues 21–26, or involves residues that make contacts with some of the helix residues.

The equilibrium between these apoenzyme conformers has no physiological significance since at intracellular concentrations of NADPH, and with the established mechanistic pathway, only a vanishingly small fraction of DHFR is present as apoenzyme, either in the cell, or under assay conditions. The kinetics of ligand binding to substrate–enzyme or product–enzyme complexes of wild type (wt) vertebrate enzymes has not provided evidence that these complexes exist in more than one conformational form, although it may occur for some variants of human DHFR (20).

B. BINDING OF INHIBITORS

1. Binding of Trimethoprim and Methotrexate

Trimethoprim and methotrexate are the two most widely investigated inhibitors of DHFR, in part because of their clinical use: trimethoprim against bacterial infections, and methotrexate in cancer therapy. Trimethoprim [2,4-diamino-5(3,4,5-trimethoxybenzyl)-pyrimidine] binds much more tightly to bacterial DHFR than to eukaryotic DHFRs. Table VIII shows that K_i values for trimethoprim with vertebrate DHFRs are 1–4 μM, whereas it is 80 pM for DHFR from *E. coli* at pH 7.65 and 20° (30). The K_i for *P. carinii* is intermediate, as perhaps might be expected for a fungal DHFR.

Methotrexate is a powerful inhibitor of both prokaryotic and eukaryotic DHFRs. Most investigations have found that for the verte-

TABLE VIII

Dissociation Constants (K_i or K_d) for Trimethoprim and Methotrexate Binding to Vertebrate DHFRs

Inhibitor	DHFR Source	K_i (nM)	Reference
Trimethoprim	Human	960	43
	Human	1000	33
	Mouse	3500	44
	Chicken	3530	34
	P. carinii	152	24
Methotrexate	Human	0.0034	43
	Human	0.15[a]	45
	Human	0.11	46
	Mouse	0.004	44
	Chicken	2.8[b]	21
	Chicken	0.009	34
	P. carinii	0.0081	24

[a] The parameter K_d for a ternary complex with enzyme. NADPH.

[b] The parameter K_d for a binary complex.

brate DHFRs, K_i is 2–8 pM (Table VIII). The value for the human enzyme was obtained from a very extensive study that thoroughly validated both the technique and the final result. Where greater values have been found, it appears likely that experimental problems have been responsible for the high values.

2. Isomerization of Ternary Inhibitor Complexes

The DHFR.NADPH.MTX complex that is initially formed undergoes isomerization, presumably involving a conformation change of some kind, to a form of the complex from which MTX can not dissociate at a significant rate, according to Scheme 1. Evidence for this is provided by the observation that when a mixture of MTX and H_2folate are added to a solution of buffer, NADPH, and a high concentration of human DHFR, the inhibition progressively develops over a period of seconds, and one of the phases of this development is due to the isomerization (43). However, at a low concentration of enzyme and MTX, and a high concentration of H_2folate, there is a slow development of inhibition (lasting for many minutes) due primarily to a slow association of inhibitor with enzyme.NADPH as the small amount of MTX competes with H_2folate for the low concentration of enzyme.NADPH. It is this competition process that accounts for so-called "slow binders" (47, 48) and not, as the latter term suggests, a low k_{on} for the inhibitor, nor a very low rate constant for the isomerization of the initial complex ($t_{1/2}$ is of the order of minutes).

Referring to Scheme 1, the values of rate constants for the human enzyme at pH 7.65 and 20° are (43) k_{on}, $1 \times 10^8 \ M^{-1} \ s^{-1}$; k_{off}, 0.02 s^{-1}; k_{iso}, 0.4 s^{-1}; $k_{r,iso}$, 0.007 s^{-1}. Thus the initial binding is quite tight (k_{off}/k_{on} 210 pM), but is increased 60-fold by the isomerization.

$$\text{MTX} + \text{H}^+$$

$$\text{DHFR. NADPH} \underset{k_{off}}{\overset{k_{on}}{\rightleftharpoons}} \text{DHFR.NADPH.MTXH}^+ \underset{k_{r,iso}}{\overset{k_{iso}}{\rightleftharpoons}} \textit{DHFR.NADPH.MTXH}^+$$

Scheme 1.   Binding of methotrexate (MTX) to the NADPH binary complex with human DHFR with concomitant protonation, and subsequent isomerization of the ternary complex. Italics indicate a conformation from which dissociation of MTX is negligible.

This result is in contrast to *E. coli* DHFR for which isomerization of the ternary MTX complex produces negligible increase in MTX binding (30).

The true K_i, or ternary K_d, for a ligand like MTX for which binding is enhanced by isomerization of the initial complex, is given by $k_{off}k_{r,iso}/[k_{on}(k_{iso} + k_{r,iso})]$. Under many circumstances K_i is approximately equal to $k_{off,app}/k_{on}$, where $k_{off,app}$ is determined by the competition method. Consequently, close agreement between k_{off} as determined by kinetics of binding and $k_{off,app}$ is an indication that there is no isomerization of the initial ligand complex. This is the case for complexes of physiological ligands with human and mouse DHFRs (31, 32), so that the evidence indicates that these complexes do not undergo any conformational change that affects binding.

The structural nature of the conformational change undergone by the initial hDHFR.NADPH.MTX complex is unknown. In the final configuration, MTX has its pteridine ring flipped over as compared with bound folate, and is protonated when bound to human DHFR (22), bovine DHFR (23), and chicken DHFR (21). This is illustrated by Scheme 2. Taira et al. (49) postulated that MTX binds initially in a "folate-like" conformation, and that in the isomerization of the

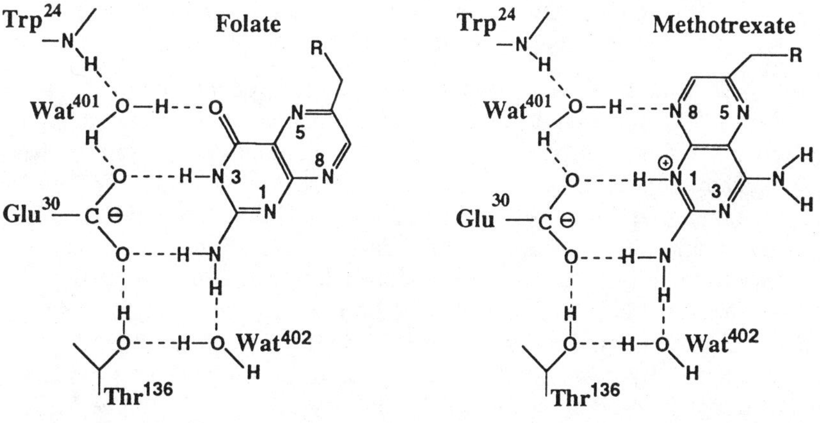

Scheme 2. Diagrammatic representation of the interaction of the pteridine rings of folate and methotrexate with active site residues. Hydrogen bonds are represented by broken lines. The R represents the *p*-aminobenzoylglutamate moiety.

ternary MTX complex the pteridine ring rotates 180° to the final orientation seen in the crystal structure. However, this seems unlikely since protonation of the pteridine ring is simultaneous with binding, and no further change in UV absorbance accompanies isomerization (50). It is improbable that MTX would be protonated when bound in the folate-like orientation, but if it were, the charge would probably reside primarily on N3. Tautomerization to put the positive charge on N1 as the ring flipped would be expected to be accompanied by a change in UV spectrum.

The weak binding of trimethoprim to vertebrate DHFRs compared with the binding to *E. coli* DHFR is in part explained by the fact that the initial complex does not undergo an isomerization (34, 43), but the value of k_{off}/k_{on} is also several thousandfold lower for the bacterial enzyme than for the vertebrate enzyme. As previously indicated (Section II.B.8), this weaker binding of trimethoprim to the vertebrate DHFRs is, at least in part, due to the 1.5–2.0 Å greater width of the active site cleft. Since Phe[31] (or Tyr[31]) in the active site of vertebrate DHFRs corresponds to a Leu in most bacterial DHFRs, it seemed possible that the Phe[31] to Leu mutation in human DHFR would confer tighter trimethoprim binding, but this is not the case (33).

3. Binding of 7-Hydroxymethotrexate, Methotrexate Polyglutamates, and Other Inhibitors

The binding of 7-hydroxymethotrexate is of clinical interest since during, and immediately after, treatment of cancer patients with high dose MTX, a high concentration of this metabolite is present in the plasma of the patient. 7-Hydroxymethotrexate has a K_i of 8.9 nM for human DHFR at pH 7.65 and 20° (43). This is 2600 times higher than for MTX. Contrary to some earlier suggestions, the K_i is not lowered at all by the presence of four additional glutamate residues on the side chain of 7-hydroxymethotrexate. The addition of four additional glutamate residues to MTX, the parent compound, decreases K_i by a factor of 2.5 to give a K_i of 1.4 pM. This modest decrease is to be anticipated from the crystal structure of the MTX complex (Fig. 5), from which it may be seen that when MTX polyglutamates bind, the polyglutamate "tail" must be entirely outside the binding pocket. A similar small increase in binding due to additional glutamate residues has been reported by others (46).

Dissociation constants for the binding of other inhibitors to highly purified eukaryotic DHFRs are collected in Table IX. Pyrimethamine is of interest because it is widely used for the treatment of malaria. It is much more tightly bound to these DHFRs than is trimethoprim (Table VIII), and less tightly bound to *P. carinii* DHFR than to the vertebrate enzymes. Piritrexim, edatrexate, and trimetrexate have been investigated as possible oncolytic agents. Epiroprim is of interest because of the relatively low K_i for *P. carinii* DHFR, compared with that for the human enzyme, a result suggesting that it may be useful in the therapy of this opportunistic pathogen

TABLE IX

Inhibition Constants (nM) for Other Inhibitors of Eukaryotic DHFRs

| Inhibitor | Source of DHFR | | |
	Human	Chicken	*P. carinii*
Pyrimethamine	1.15 ± 0.06[a]	6.2 ± 0.6[b]	9.7 ± 0.3[a]
Piritrexim[c]	0.0077 ± 0.0012[a]		0.143 ± 0.015[a]
Epiroprim[d]	59.1 ± 4.2[a]		17.4 ± 1.2[a]
Edatrexate[e]	0.001 ± 0.0001[a]		0.0039 ± 0.0001[a]
Trimetrexate[f]	0.0019 ± 0.0003[a]		0.71 ± 0.04[a]
DAMP[g]		285 ± 17[b]	
Triazine[h]		0.4 ± 0.04[b]	
10-FormylH$_2$folate	17 ± 1[i]		
10-FormylH$_2$folate.Glu$_5$	6.4 ± 0.1[i]		
5-Deazafolate	0.7[j]		

[a] Reference 24: at pH 7.0.

[b] Reference 34: at pH 7.4.

[c] 2,4-Diamino-6-(2,5-dimethoxybenzyl)-5-methylpyrido[2,3-*d*]-pyrimidine.

[d] 2,4-Diamino-5(3,5-dimethoxy-4-pyrrol-1-ylbenzyl)pyrimidine.

[e] 10-Ethyl-10-deazaaminopterin.

[f] 2,4-Diamino-5-methyl-6-[(3,4,5-trimethoxyanilino)methyl]quinazoline.

[g] 2,4-Diamino-6,7-dimethylpteridine.

[h] 4,6-Diamino-1,2-dihydro-2,2-dimethyl-1-(4-phenyl-thiomethylphenyl)-1,3,5-triazine.

[i] Reference 46: at pH 7.4.

[j] Reference 22: at pH 7.65.

that causes pneumonia in immune-compromized patients. The triazine is unusual in that it has been reported to show extremely high positive cooperativity (48,000) for chicken DHFR. This value may be compared with cooperativities for methotrexate of 480 for human DHFR (50), and 120 for chicken DHFR (34). 10-FormylH$_2$folate and its polyglutamates have been proposed as physiological inhibitors of vertebrate DHFR under conditions of thymidylate synthase inhibition. 5-Deazafolate is significantly more inhibitory than folate itself, which has a K_i of 7 nM for human DHFR (22). The complex of 5-deazafolate with human DHFR has been studied by X-ray crystallography by Davies et al. (18), who regard it as a model for the transition state for folate reduction.

C.   ROLE OF SPECIFIC SIDE CHAINS IN LIGAND BINDING

1.   Role of Glu^{30}

This residue is rigorously conserved in all eukaryotic DHFRs, is an Asp in all bacterial DHFRs, and is Asp or Glu in other DHFRs. The side chain carboxyl of Glu^{30} in human DHFR is hydrogen bonded to N3 and the 2-amino group of bound folate (Fig. 2). A similar interaction is made between Glu^{30} of chicken DHFR and bound biopterin (Fig. 4). It might therefore be reasonably assumed that substitution of another residue for Glu^{30} would greatly influence ligand binding. Thillet et al. (32) examined the E30D variant of mouse DHFR (i.e., mouse DHFR with Asp^{30} instead of Glu^{30}) and found relatively small changes in k_{off} for physiological ligands. In most cases K_d was also unchanged or decreased a little. This result probably indicates that the Asp side chain is able to make similar interactions with bound pterins to those made by Glu^{30}, but there is no crystallographic evidence that this is the case. The E30D variant of human DHFR has also been prepared and in binary complexes K_d for NADPH is similar to that for wt, but that for H$_2$folate is increased about three-fold (51). The E30Q variant of human DHFR also shows little change in binding of NADPH. Although K_d for H$_2$folate has not been determined, the high K_m, 100 μM, suggests that binding of H$_2$folate is weakened for the E30Q variant (51). The fact that k_{off} for folate, H$_2$folate, and H$_4$folate dissociation from human DHFR increases 6–30-fold as the pH is raised from 5 to 7 (52) may well be due to an effect on the interaction of Glu^{30} with these ligands.

The carboxyl group of Glu[30] also interacts with bound methotrexate forming an ionic linkage and hydrogen bonds with the protonated N1 and the 2-amino group of MTX (Fig. 5). Although experimental data are not yet available to estimate the binding energy contributed by the ionic bond to the binding of methotrexate to a vertebrate DHFR, in the case of the *E. coli* enzyme it is 5.2 kcal/mol for the ternary complex, and 3.9 kcal/mol for the binary complex (30). This energy of the ionic bond accounts for most of the additional binding energy for methotrexate as compared with folate.

2. Role of Hydrophobic Residues Lining the Active Site

As indicated in Section II.B, the crystallographic data indicate that in the complex of folate with human DHFR the pterin ring is surrounded by a number of hydrophobic residues: Phe[34], Phe[31], Leu[22], and Ile[7]. The pABA moiety is similarly encompassed by Phe[34], Ile[60], Asn[64], and Leu[67] (Fig. 2). The hydrophobic and, in some cases, van der Waals interactions with the bound ligand make an important contribution to binding energy. These residues are highly conserved. Leu[67] is invariant in all known sequences, Phe[34] is invariant in eukaryotic sequences and is present in most other sequences, Ile[60] is invariant in eukaryotic sequences, and Asn[64] is invariant in vertebrate sequences. Where these residues are not invariant replacements are almost always highly conservative.

A first indication of the role of Phe[31] in ligand binding was provided by an examination of the F31L variant of human DHFR (20). This substitution of Leu for Phe[31] causes small decreases in k_{on} for folate, H_2folate, and H_4folate, but the significant effects are on k_{off} for complexes of these pterins, with decreases by factors of 7–500. Although the decrease in side chain bulk might be expected to decrease hydrophobic interaction, with a corresponding *increase* in k_{off}, this appears not to be the case. Further analysis of the role of Phe[31] in ligand binding was obtained by constructing variants of recombinant human DHFR in which this residue was replaced by Val, Thr, Ser, Ala, or Gly (50, 53, 54). For all these variants K_d for binary complexes with folate were significantly decreased (to 12–78% of the wild-type value) and K_d for the H_2folate binary complex was considerably decreased for all variants (to 1.5–5.4% of the wild-type value). For the F31G variant, as in the case of the F31L variant, this was due to decreases in k_{off}. For the binary complex of H_4folate with the F31G variant k_{off} was unaltered, but those for ternary complexes involving H_4folate were significantly decreased.

When the binding of methotrexate to these variants was examined, K_i was unchanged for the Leu, Val, and Thr variants, but was increased 70–100-fold for the Ser, Ala, and Gly variants. This change is the opposite to that seen for folate and its reduced forms, but the mechanism is different. Methotrexate binding to the F31G variant was examined in detail, and k_{off} was found to be *decreased* as in the case of the folates. The increase in K_i is due to a decrease in k_{on} (by a factor of 5), and elimination of isomerization of the initial complex.

How these side chain substitutions produce the observed kinetic effects is unclear. Crystallographic results for the F31S, F31A, and F31G variants (50, 55) indicate that there are no major changes in protein backbone conformation, in side chain positions, or in the position of bound inhibitor. An additional bound water molecule occupies the site vacated by the phenyl moiety of the side chain at position 31. It is perhaps the latter that is the critical factor in decreasing the off rate for bound ligands.

In the complex of folate with human DHFR, Phe[34] partially covers the face of the pteridine ring on the side opposite the nicotinamide binding site, and the plane of the phenyl ring is inclined at an angle of about 45° to the plane of the pteridine ring (Fig. 2). The edge of the phenyl ring is in van der Waals contact with the pteridine ring. It interacts similarly with the pteridine ring of bound methotrexate (19). Replacement of Phe[34] by Ser, Thr, or Ala causes large increases in K_d for the binary complexes with H_2folate, and for these variants and for F31I and F31V there are large increases in K_m for H_2folate (54, 56, 57). Much larger decreases in affinity for methotrexate also result from these mutations and in the case of the F34S variant the increase is by a factor of 60,000. In the case of the F34A variant the weak affinity for MTX is due in part to a decrease in k_{on} by factors of 30 and 40 for binding to the enzyme and to its NADPH complex, respectively. However, the major cause of the decrease in affinity is a large increase in k_{off}: 253- and 460-fold for the binary and ternary complexes, respectively. Crystal structures of these variants are not yet available.

Leu[22] of human DHFR has its side chain methyl groups beyond the edge of the pteridine ring of bound folate (Fig. 2) or MTX and below the plane of the pteridine ring, on the same side as the nicotinamide ring of bound NADPH. The closest contacts of these methyl groups with MTX are with C6, C7, and N8, from which the shortest separations are 3.60, 4.07, and 4.41 Å, respectively (58). Replacement of Leu[22] by Tyr, Trp, or Arg *increases* the binding of H_2folate and replacement by Phe causes a modest decrease. All four variants bind H_4folate less tightly, but only by factors of 6-23. The most remarkable effect of these substitutions is on the binding of inhibitors. In the ternary DHFR.NADPH.MTX complexes MTX binding is decreased by factors of 740-28,000 (58), and this is due to large increases in k_{off} values. The binding of aminopterin, edatrexate (10-deaza-10-ethylaminopterin), piritrexim, and trimetrexate are also decreased by factors of 33-960,000, depending on the inhibitor and the variant, with the greatest decrease in binding that for trimetrexate binding to the Arg[22] variant.

The decrease in the binding of MTX has been studied in some detail, and for all four variants the major cause is a large increase in k_{off} for MTX dissociation from the ternary complex (2300- to 4100-fold). There are also contributions from small decreases in k_{on}, but not from changes in K_{iso}, which actually increases for each variant except Arg[22].

A number of these complexes have been crystallized and subjected to structural analysis. The substitutions cause little perturbation of the backbone or side chain structure, and the largest shifts in the backbone are in flexible loops some distance from the active site. The side chains of Tyr[22] and Arg[22] adopt low probability conformations in the ternary complexes in order to avoid unfavorable contacts with the inhibitor and NADPH. There seems to be little loss of interactions between bound inhibitors and the protein due to the substitutions at position 22, except in the case of the Arg[22] variant, and if the pteridine ring exits between Phe[31] and residue 22 that path seems no more open in the variants than in the wt enzyme structure. The structural basis for the increase in k_{off} is therefore unclear. However, the activation energy for MTX dissociation may be lowered by the adoption of lower energy conformations by Tyr[22] and Arg[22] when the pteridine site is emptied.

3. Role of Arg[70]

In complexes of human DHFR the side chain guanidinium group of this residue is ionically bonded and hydrogen bonded to the α-carboxyl group of bound folate (Fig. 2) or bound methotrexate (19). This residue is conserved not only in the nine eukaryotic DHFRs but also in five protozoan DHFRs, six DHFRs coded for by viruses, plasmids, and transposons, and in seven out of eight bacterial DHFRs, so that this interaction is presumably important in binding of folates (and analogs that retain the pABGA moiety). Replacement of Arg^{70} by a Lys causes a considerable decrease in affinity for methotrexate and H_2folate at pH 7.5 and a six- to eightfold greater loss of binding energy at pH 8.5, when the lysine side chain is unprotonated (59).

4.   Role of Lys^{54}

In the crystal structure of the ternary chicken DHFR.NADP.biopterin complex (15), and of the chicken DHFR.NADPH.phenyltriazine complex (13) the 2'-phosphate of the bound dinucleotide makes a charge mediated hydrogen bond with Lys^{54}. There is a similar charge and hydrogen-bond interaction in the chicken DHFR.NADPH binary complex (27). This interaction must contribute significantly to the energy for binding of NADPH and NADP.

In confirmation of this, the mutation producing the $Lys^{54} \rightarrow$ Gln substitution in human DHFR decreases NADP binding to the apoenzyme by a factor of 126, due to both a decreased k_{on} and increased k_{off} (60). The parameter K_m for NADPH is increased 58-fold, and the K_m for NADH is increased to a lesser extent, so that for the variant the ratio of K_m values is 4.7 compared with 69 for wt. The parameter K_i for NADP is also increased 58-fold. Thus Lys^{54} is important in determining the specificity for NADPH. However, k_{cat} is increased 2.5-fold by the mutation. This increase is presumably caused by faster dissociation of NADP from the ternary product complex, and all of the flux probably goes through this step.

D.   MOTION OF BOUND LIGANDS

A ligand that is tightly bound in the active site is not necessarily immobilized. An indication of the amount of movement of atoms of a bound ligand can be obtained from the thermal factors for the relevant atoms in X-ray crystallographic analysis. In general the examination of these factors for published structures shows that the pteridine ring of folate bound to human DHFR is held rather rigidly with average thermal factor for its atoms similar to or lower than the average for protein backbone atoms in the β sheet (17, 61). The same is true for the 2,4-diaminopyrimidine ring of trimethoprim bound to chicken DHFR (61). Atoms in the benzene ring of either ligand are less constrained, and there is still more motion for atoms of the glutamate moiety, especially the γ carboxyl and two adjacent carbon atoms.

Another approach to this question is the use of ESR (electron spin resonance) to examine the mobility of a nitroxide probe attached to various inhibitors (61). When the probe is attached to the 6- position of 2,4-diaminopyrimidine or 2,4-diaminopteridine, so that the probe is presumably held within the active site cavity, it is highly immobilized when the ligand is bound to bovine or human DHFR, with correlation times, τ_c, of 4–20 ns. When the probe is attached to the α carboxyl of methotrexate bound to these enzymes immobilization is almost as great, but there is negligible restriction of the motion of a probe attached to the γ carboxyl.

IV. Kinetic Scheme

This section deals with the catalytic pathway in the sense of the order of addition of substrates and the release of products. The results of steady-state kinetic investigations will be considered first, and then the considerably greater insights resulting from the use of transient state kinetic techniques will be described. As it turns out, human DHFR has a branched pathway, and changes in conditions that alter the flux through the alternate pathways alter the steady-state kinetics in ways that are hard to interpret if the transient state kinetic data are not available.

A. STEADY-STATE KINETIC RESULTS

The available steady-state parameters obtained with highly purified enzymes from eukaryotic sources are summarized in Table X. It may be seen that, although there are exceptions, vertebrate DHFRs have lower K_ms for H_2folate, and lower values of k_{cat}, than DHFR from lower eukaryotes. NADP has been reported to be inhibitory for several of the vertebrate enzymes. Although there are few reports of investigation of the kinetic pathway for these enzymes

TABLE X

Steady State Kinetic Parameters for Eukaryotic DHFRs

| | K_m (μM) | | | | Conditions | | |
| | | | $k_{cat}{}^a$ | $K_i{}^b$ | | Temp. | |
Source	H_2folate	NADPH	(s^{-1})	(μM)	pH	(°C)	Reference
P. carinii	2.5	8.9	136		7.0	20	24
C. albicans	2.5	3.3	183		7.0		62
C. neoformans	0.44	1.7	17.5		7.0	25	4
Chicken[c]	~0.15	1.8	5.4	2.4	7.4	28	63
Bovine	1.9				6.0	20	61
Bovine			9.6		7.0	28	64
Murine[d]	0.9	6.2, 49	36		7.0	25	32
Murine	0.13	0.32	5.9		7.65	20	64a
Porcine	0.74	3.22	12.8	2.2	7.6	30	8
Human	0.12	0.16, 4.2	12.5	0.21	7.65	20	31
Rat	0.17	0.72	2.32		6.5	21	65

[a] With H_2folate as substrate.

[b] For NADP.

[c] Since v^{-1} is not a linear function of s^{-1}, estimation of parameters is difficult.

[d] In 0.8 M NaCl.

by steady-state methods (i.e., determination of steady-state initial rates at various concentrations of substrates and products), those that have appeared have been contrary to expectation. Thus there are two reports (24, 62) of initial velocity patterns for DHFRs like those expected for a Ping Pong mechanism, though this is not an expected mechanism for a dehydrogenase. These and other observations can be understood in terms of the mechanistic scheme for these enzymes based on transient state kinetic results. Verification of the resulting kinetic scheme can be obtained by computer simulation (based on the rate constants involved) of the following: transient state kinetic behavior; steady-state behavior, including K_m values, k_{cat}, and product inhibition or activation; and reaction progress curves.

B. KINETIC SCHEMES BASED ON TRANSIENT STATE KINETIC RESULTS

A quantitative description of the contribution of all possible kinetic pathways to the flux through an enzyme reaction can be ob-

tained from the rate constants for all possible association and dissociation reactions of substrates and products and for the forward and
reverse chemical reactions. This method was first applied to DHFR
by Benkovic and co-workers (66) who described the scheme for *E.
coli* DHFR. The rate constants in the forward direction for each step
in a particular pathway together determine the quantitative importance of that pathway.

1. Kinetic Scheme for Human DHFR

The results of the method applied to human DHFR are shown in
Scheme 3 (31). It may be seen that the ternary substrate complex,
E.NADPH.H$_2$folate, is rapidly converted to the ternary product
complex, E.NADP.H$_4$folate. This chemical transformation step,
with rate constant 1360 s^{-1}, is discussed in detail in Section V. After
that step, either of the products can dissociate to leave a binary

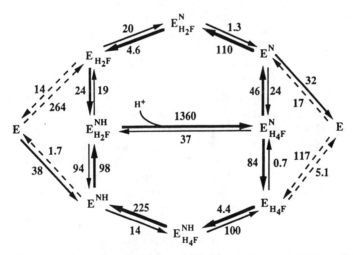

Scheme 3. Kinetic scheme for human dihydrofolate reductase at 20°C, in MATS
buffer, pH 7.65 (31). E = human DHFR, NH = NADPH, N = NADP, H$_2$F =
H$_2$folate, H$_4$F = H$_4$folate. Association and dissociation rate constants have the units
μM^{-1} s^{-1} and s^{-1}, respectively. The steps in the major pathways in the presence
of saturating substrate concentrations are depicted by *bold solid arrows*. When
NADPH is saturating but the concentration of H$_2$folate is low, significant flux also
occurs through the path E.N, E, E.NH (*thin arrows*). *Broken arrows* indicate less
significant pathways.

complex, and the rates of these two dissociations are not very different (k_{off} = 46 and 84 s^{-1} for H$_4$folate and NADP, respectively). This
results in a branched pathway, and at steady state, in the presence of
saturating substrate concentrations, the flux through the E.H$_4$folate
pathway (the lower loop in Scheme 3) is 65% of the total. In this
pathway the dissociation of H$_4$folate from the binary complex is
slow (k_{off} = 5.1 s^{-1}) and at concentrations of NADPH under assay
conditions, or under intracellular conditions, NADPH rapidly binds
to form the E.NADPH.H$_4$folate complex. As discussed in Section
III.A.3, there is very strong negative cooperativity of the binding of
these ligands, so that H$_4$folate dissociates rapidly from this complex
(k_{off} = 225 s^{-1}). Thereafter, H$_2$folate binds to complete the cycle.
This final step in this branch of the pathway is very rapid if the
H$_2$folate concentration is saturating, but is slow enough at intracellular concentrations of H$_2$folate to limit the rate of turnover.

Considering next the branch through E.NADP, the course of this
pathway depends on the concentration of H$_2$folate. If H$_2$folate is
saturating it rapidly binds to give the E.NADP.H$_2$folate complex.
This complex has significant positive cooperativity, because dissociation of NADP is slow (k_{off} = 4.6 s^{-1}), and this abortive ternary
complex accumulates during several cycles of product formation before it finally reaches a steady-state level of 87% of the total enzyme
present. As this abortive complex accumulates, hysteresis occurs,
that is, there is a continuous change in the overall rate of the reaction
as the flux shifts increasingly from the faster lower loop in Scheme
3 to the slower upper loop. If, however, the H$_2$folate concentration
is low, then most of the E.NADP complex dissociates to give apoenzyme, to which NADPH rapidly binds followed by H$_2$folate binding.
Thus at high H$_2$folate concentrations the rate of this pathway is limited by the dissociation of NADP from the abortive ternary complex,
but at low concentrations the limitation is the rate of H$_2$folate
binding.

This scheme correctly predicts all aspects of transient state and
steady-state kinetics that have been observed. It quantitatively predicts the observed inhibition (at saturating H$_2$folate) by NADP and
activation by H$_4$folate, since these products serve to drive flux more
into the slower upper pathway or faster lower pathway, respectively.
It correctly predicts k_{cat}, and K_m values, including the two observed
for NADPH, which correspond to the addition of NADPH to E.H$_4$-

folate and to E.H$_2$folate, respectively. It also predicts the transient state hysteresis observed.

2. Kinetic Scheme for Murine DHFR

The kinetic scheme for the mouse DHFR has been reported by Thillet et al. (32). Since it is based on rate constants obtained in the presence of 0.8 M NaCl, and since some of the rate constants are significantly different at physiological ionic strength (Table X), it is uncertain whether this is the scheme that is operational under the latter conditions.

This scheme resembles that for human DHFR in that the chemical transformation step is much faster than the dissociation of ligands at neutral pH. However, H$_4$folate dissociates much more slowly than NADP (k_{off} values 18 and 450 s^{-1}, respectively), so that there is no flux through the upper loop of Scheme 3. Binding of NADPH to the E.H$_4$folate complex shows some negative cooperativity, but much less than in the case of the human enzyme, so that dissociation of H$_4$folate from the E.NADPH.H$_4$folate complex has a k_{off} of 40 s^{-1}, compared with 15 s^{-1} for the binary H$_4$folate complex. Consequently, at low NADPH concentrations virtually all of the flux is via E.H$_4$folate, E, and E.H$_2$folate, and the apparent K_m for NADPH is 6.2 μM. In the high range of NADPH concentrations most of the flux is through the ternary complex formed by NADPH binding before H$_4$folate dissociation and the apparent K_m is 49 μM.

The key differences between the mouse DHFR scheme and Scheme 3 are (a) a much higher k_{off} for NADP dissociation from the ternary product complex, (b) a somewhat lower k_{off} for H$_4$folate from that complex, and (c) a much smaller negative cooperativity for the E.NADPH.H$_4$folate complex. Since there is no branched pathway like that in operation for the human enzyme, there is no hysteresis during the first few turnovers like that observed for the human enzyme, which is due to redistribution of flux through the two main pathways. In fact the very high ionic strength selected for examination of the mouse kinetics was chosen precisely because no hysteresis occurs under these conditions (32). However, at lower ionic strength such hyseresis does occur for the mouse enzyme also (32), presumably because k_{off} for NADP dissociation from the ternary complex is much lower under these conditions, resulting in a branched pathway that probably resembles that in Scheme 3 quite closely.

3. Kinetic Scheme for P. carinii DHFR

The kinetic scheme for *P. carinii* DHFR (24) is shown in Scheme 4. In contrast to human DHFR, the dissociation of NADP from the ternary product complex, $E.NADP.H_4$folate, is extremely rapid with a rate constant too high to be measured in the stopped flow fluorimeter. As a result no significant flux goes through the upper pathway in Scheme 4. The dissociation of H_4folate from its binary complex is even slower that in the case of human DHFR, so that there is negligible formation of apoenzyme at any time during the cycle. When NADPH binds to give $E.NADPH.H_4$folate the negative cooperativity is high, so that H_4folate rapidly dissociates (k_{off} = 216 s^{-1}). The combination of the moderately high rate constant for the chemical transformation step (402 s^{-1}) and the high k_{off} values for the products results in a considerably higher k_{cat} than for the vertebrate

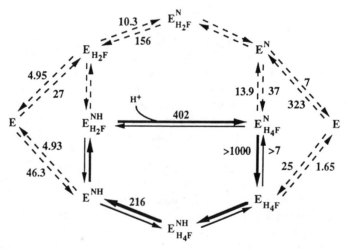

Scheme 4.   Kinetic scheme for DHFR from *P. carinii* at 20°C, in MATS buffer at pH 7.0 (24). Abbreviations and units for rate constants as in Scheme 3. The significant pathway is indicated by *solid* arrows, *bold* in the forward direction. Abbreviations and units for rate constants as in Scheme 3.

enzymes. It was found that the pathway in Scheme 4 is described by a steady-state rate equation of the form $v/[E] = k_{cat}AB/(K_aB + K_bA + AB)$, which is also the equation for a classical Ping Pong mechanism. The fact that the steady-state results for *C. albicans* DHFR also fit to this equation (62) and that k_{cat} is high suggests that this enzyme has a kinetic scheme similar to that of *P. carinii* DHFR.

4. Effect of Mutations on the Kinetic Scheme

Mutations that change the rate constant for the chemical transformation step, or for one or more product dissociation steps, or both, can drastically alter the kinetic scheme for the enzyme, with consequent effects on the steady-state kinetic constants that can be quite considerable.

A striking illustration of this is the effect of mutations causing the substitution of other amino acids for Phe[31] in human DHFR. The position of this residue in the catalytic site and its role in substrate binding has been discussed previously (Section II.C.2). The rather conservative replacement of this residue by Leu causes marked changes in many of the kinetic properties (Table XI). The Michaelis

TABLE XI
Kinetic Behavior of Variants of Human DHFR

Variant enzyme	K_m (μM) NADPH	K_m (μM) H$_2$folate	k_{cat} (s^{-1})	k_{chem} (s^{-1})	Transient state hysteresis	Reference
wt	0.16, 4.2	0.12	12.5	1360	Yes	31
E30D	4.4	0.30	23			51
E30Q	0.4	100	0.30			51
L22Y	0.91	0.53	6.1	9.1	No	58
L22F	2.34	0.33	24	24	No	58
L22W	1.35	0.42	4.2	3.3	No	58
L22R	1.2	1.6	0.045	3.0	No	58
W24F	3.4	2.5	40	89	No	42
F31L	21.9	0.77	22.2	520	No	20
F31G	0.77	0.43	11.3	27	No	50, 53
F34A	0.74	36[a]	8.4[b]	6.8	No	56, 57

[a] K_d for H$_2$folate dissociation from E.NADPH.H$_2$folate: substrate inhibition by H$_2$folate makes direct determination of K_m impossible.

[b] Maximum rate through the faster pathway.

constants for NADPH and H_2folate are both increased, but although the rate of the chemical transformation step is decreased considerably, k_{cat} is doubled. This variant does not show hysteresis during the first few turn overs. The explanation of these changes is found in changes of k_{off} and k_{on} for substrates and products. The parameter k_{off} for NADP *increases* by a factor of 1.6, whereas k_{off} for H_4folate *decreases* by a factor of 46. The result is that for this variant of human DHFR, flux is entirely via the E.H_4folate complex. The dissociation of H_4folate from this complex is very slow (k_{off} = 0.15 s^{-1}), and negative cooperativity is high for the E.NADPH.H_4folate complex, so all the flux is by way of this complex, that is, solely by the lower bold arrow pathway in Scheme 3. Consequently there is only one K_m for NADPH (and it is high), and since there is no branched pathway, there is no hysteresis during the first few turnovers. Although the rate constant for the chemical transformation k_{chem} is decreased, it is still high compared with the rate of H_4folate release, and it is the latter that primarily limits the overall reaction rate.

Somewhat similar effects are produced by the mutation replacing Phe[31] of human DHFR by Gly (50, 53), but the effects on rate constants are less (Table XI), despite the much greater decrease in the bulk of the side chain. In this case, however, k_{cat} (which is the same as for wild type) is limited by the much lower k_{chem}, and by the rate of NADP dissociation.

The four variants with substitutions for Leu[22] have even lower values of k_{chem} and for three of them this step becomes rate limiting so that hysteresis disappears. The Arg[22] variant has, however, a value of k_{cat} that is even lower than k_{chem} at pH 7.65. Although k_{off} values for products have not been measured, one or both of these must be rate limiting.

The effect on kinetic behavior of replacing the strictly conserved residue Trp[24] by Phe is more subtle (42). Although k_{chem} is greatly *decreased* by this substitution (Table XI), k_{cat} is *increased* due to increased k_{off} values for product dissociation. In particular, k_{off} for NADP dissociation from the E.NADP.H_2folate complex, a rate-limiting step for wild-type enzyme, is increased over 10-fold and the dissociations of NADP and of H_4folate from the ternary product complex occur at almost equal rates (130 and 125 s^{-1}, respectively). Flux through both major loops (Scheme 3) is rapid and equal, and there is very little redistribution of flux through the two loops during the first few turnovers and therefore no hysteresis.

A number of variants of human DHFR that have amino acids with less bulky side chains substituted for Phe[34], show strong inhibition by H_2folate so that there is not a normal hyperbolic relationship between reaction rate and H_2folate concentration (56, 57). This substrate inhibition indicates the existence of alternate pathways each with its own "K_m" and "k_{cat}", and values for these can only be approximated indirectly (Table XI). The substrate inhibition is due to a rapid dissociation of H_4folate from the ternary product complex to give the E.NADP binary complex, and there is insignificant flux through the lower loop in Scheme 3. At low concentrations of H_2folate, NADP dissociates from the enzyme followed by binding of NADPH to the liberated apoenzyme, and finally H_2folate binding to form the ternary substrate complex. At higher H_2folate concentrations another pathway is followed, however, with binding of H_2folate to E.NADP before the latter can dissociate, and slow release of NADP from the ternary complex (i.e., the upper loop in Scheme 3). The increasing diversion of flux through this slower pathway as H_2folate concentration is increased accounts for the substrate inhibition. Similar kinetic behavior has been found for the L22F variant of human DHFR (58).

V. The Chemical Transformation Step: Hydride and Proton Transfer

The conversion of enzyme-bound substrates to enzyme-bound products by DHFR is often referred to as "hydride transfer." However, hydride transfer is only one of the aspects of the chemical transformation step. Proton transfer is equally a part of it and, as will be discussed, a conformational "flexing" of the ternary complex may play an essential role also. Each of these aspects will be discussed in turn.

Rate constants for vertebrate DHFRs for the chemical transformation step, measured during the conversion of enzyme-bound substrates to enzyme-bound products, cover a wide range of values (Table XII), with that for the human enzyme 76 times that for the chicken enzyme. The significant effect of high ionic strength on k_{chem} is evident from the two sets of results with mouse DHFR. As dis-

TABLE XII

Rate Constants for the Chemical Transformation, Hydride Transfer, and the
Isomerization for Vertebrate DHFRs

DHFR Source	Measured Constants (s^{-1})				$k_{hyd}{}^a$ (pH Independent)	Reference
	k_{chem}	k_{hyd}	k_{iso}	pH		
Human	1360	3,000	2,200	7.65	109,500	67
Mouse	>700			7.65		67
Bovine	432	1,000	675	7.65	36,500	67
Mouse	520^b	520^b	≫520	7.5	$13,100^b$	32
P. carinii	402	402	≫402	7.0	3,600	24
Chicken	18	30	27	7.65	1,100	67
E. coli^c	50	50	235	7.65	706	67

^{a} Calculated assuming an apparent pK_a of 6.1 for pH dependence of k_{hyd}, except in the case of E. coli DHFR for which a value of 6.5 was assumed (66).

^{b} In the presence of 0.8 M NaCl.

^{c} For comparison.

cussed more fully in Section V.C, in many cases it must be concluded that $k_{chem} \neq k_{hyd}$ because the isotope effect for k_{chem} (i.e., the ratio of the value with NADPH to that with NADPD) is less than that for k_{hyd} obtained, for example, at high pH. Calculated values for k_{hyd} are shown for the specific pH of the measurement of k_{chem}, as well as the pH independent value calculated from the pK_a for the pH dependence of k_{hyd}. The latter demonstrate even more clearly the wide range of efficiencies for hydride transfer by the vertebrate DHFRs, and the fact that k_{hyd} for E. coli DHFR is lower than any of the vertebrate DHFR values. Calculated values of the rate constant k_{iso} for the isomerization postulated to precede hydride transfer are also given.

A. ACTIVATION OF REACTANTS BY THE ENZYME

As with any enzyme reaction, a fundamental question of great interest is, How does the enzyme do it? How does the protein catalyse the chemical transformation of the reactants to the products? In the case of DHFR a number of hypotheses have been put forward that will be briefly discussed in turn. One or more of them may have some validity, but it is not yet a question that can be considered resolved.

1. Proposed Specific Activation of NADPH as Hydride Donor

Filman et al. (68) proposed that interactions of the dihydronico-tinamide moiety of NADPH bound to *Lactobacillus casei* DHFR facilitate entry of this group into its transition state conformation. They noted that three oxygen atoms of protein residues are located approximately within the plane of the dihydronicotinamide ring, and are adjacent to its C2, C4, and C6 positions at distances about 0.2 Å, too short for ordinary van der Waals contacts. They proposed that interaction of the dipoles of these groups with the nicotinamide carbon atoms stabilizes partial positive charges on them in the transition state. This, they argued, assists conversion of the dihydronico-tinamide ring in the transition state to an electronic and conformational isomer of NADP in which N1 has a pyramidal configuration with the plane of the ring tilted in relation to the N1—C1' bond, and partial positive charges on C2, C4, and C6. The oxygen atoms adjacent to C2, C4, and C6 are the carbonyl oxygen atoms of Ile[13] and Ala[97], and the side chain oxygen of Thr[45]. The corresponding residues in vertebrate DHFR are Ile[16], Val[115], and Thr[56], respectively, whose oxygen atoms occupy approximately similar positions relative to C2, C4, and C6 of the nicotinamide rings of bound NADPH or NADP (13, 15, 19).

Filman et al. (68) suggest two other interactions that might facilitate attainment of the transition state configuration by the nicotinamide ring. The carboxamide group of NADPH bound to *L. casei* is rotated almost 180° from its most stable conformation, with the result that there is crowding between the C2 hydrogen and a hydrogen attached to N7. This crowding is said to be relieved by the tilting of the nicotinamide ring relative to the N1—C1' bond, as the transition state is entered, but it is not obvious how the crowding is relieved. Finally, a bound water molecule (Wat[439]) that is hydrogen bonded to the side chain of Ser[48] (Ser[59] in vertebrate DHFRs) is near a position from which to donate a hydrogen bond to N1 of the nicotinamide ring when the latter enters a pyramidal configuration. From the crystal structures of vertebrate DHFRs it does not seem that there is a corresponding bound water molecule that could play a similar role in their mechanism of catalysis.

2. Facilitation of Entry into the Transition State

Kraut and co-workers (69) proposed that a major mechanism by which DHFR from various sources assists the ternary substrate to change from the ground to the transition state concerns the overlapping of the binding sites for the nicotinamide and pteridine rings. The first indication that this was the case was obtained from the crystal structure of *E. coli* DHFR complexed with NADP and folate, where the separation of C4 of the nicotinamide of NADP and C7 of the pteridine ring of folate is only 3.1 Å compared with a predicted van der Waals separation of 3.6 Å. The overlap of these binding sites became much more obvious when NADPH was modeled into the crystal structure of the binary complex of folate with human DHFR (18). This hypothetical model of the ternary complex was generated by an α-carbon least-squares superposition of the 1.7-Å refined chicken DHFR.NADPH coordinates and the human DHFR coordinates. The high degree of similarity in the NADPH binding sites of the two enzymes is reflected in the rms deviation of 0.4 Å for the superpositioning of α-carbon atoms of 22 residues that contact NADPH in the chicken DHFR.NADPH structure. Of these residues, 18 contribute side chain atoms to those interactions, and only two of these 18 are not identical in the chicken and human DHFR structures and neither is near the folate binding site. A case is made that these slight differences are unlikely to introduce errors in the predicted binding of NADPH in the hypothetical model.

In the hypothetical model of the human DHFR.NADPH.folate complex all other nonspecific and hydrogen-bonding interactions seen in the chicken DHFR complex can also be formed, and there are no unfavorable contacts between the enzyme and NADPH. However, the pteridine ring protrudes approximately 1 Å into the unoccupied nicotinamide binding pocket of the human DHFR structure. This protrusion is regarded not as the result of uncertainties in the modeling, but as evidence for purposeful overlap of the binding sites. It is noteworthy that in the model the nicotinamide C4 is positioned almost directly under and 2.4 Å away from C7, the carbon of folate to which hydride must be added, while C4 of nicotinamide is 2.6 Å from C6 of folate, the carbon of H_2folate to which hydride is added.

Davies et al. (18) see special significance in this close approach of the hydride donor atom of NADPH and the hydride acceptor atom of substrate, since ab initio calculation of the transition state structure for reduction of methyleniminium cation by 1,4-dihydropyridine indicates that a 2.6-Å separation of the donor and acceptor atoms is optimal for hydride transfer (70). Similar calculations for 1,4-dihydropyridine reduction of cyclopropenium cation indicate a transition state in which the nicotinamide ring is nearly planar, with C4 only 0.1 Å out of the plane in the direction of hydride transfer (71).

Davies et al. (18) obtained what they believe is an even better model of the transition state for folate reduction by DHFR by superimposing the chicken DHFR.NADPH and human DHFR.5-deazafolate structures. The principal additional feature of this model, compared with the one based on the human DHFR.folate complex, is an inferred protonation of N8 of the bound 5-deazafolate and a hydrogen bond between N8 and the carbonyl oxygen of Ile[7]. This hydrogen bond is achieved not by reorientation of the heterocyclic ring, but by a 22° rotation of the Ile[7]—Val[8] peptide plane. This formation of a hydrogen bond is believed by Davies et al. (18) to model the protonation of N8 of folate in the transition state, and to indicate that similar hydrogen bonding of N8 of protonated folate to the Ile[7] oxygen would occur. They also conclude that a similar hydrogen bond would be formed to N8 of bound H_2folate.

McTigue et al. (15) reported that, in the crystal structure of the complex of chicken DHFR with NADP and biopterin, the separation of C4 of the nicotinamide ring and C7 of the oxidized pteridine ring is 3.3 Å instead of the predicted van der Waals separation of 3.6 Å. From a comparison of the crystal structure of this complex and that of the isomorphous chicken DHFR.NADPH structure they conclude that entry of the ground-state ternary complex into the transition state requires movement of both the reduced nicotinamide ring and the pteridine ring of the substrate. The position of the substrate pteridine ring is nearly identical in the *E. coli* DHFR.NADP.folate and chicken DHFR.NADP.biopterin ternary complexes, whereas the positions of the pteridine ring in the chicken ternary complex and in the human DHFR.folate binary complex differ by an approximately 20° rotation about an axis collinear with the C4a—N5 pteridine bond. This results in the pteridine ring of folate bound to human DHFR extending about 1 Å deeper into the catalytic site.

The driving force for movement of the nicotinamide ring of NADPH and the substrate pteridine ring to approach each other in the transition state, and thus occupy positions close to those that they have in the respective binary complexes, presumably comes from close contacts between enzyme side chains and the face of each ring opposite to that engaged in hydride transfer.

3. Proposed Activation of Pteridine as Hydride Acceptor

Bajoreth et al. (72–74) investigated the electronic redistribution that occurs in folate and H_2folate bound in the active site of DHFR from *E. coli*. To determine this migration of electron density in the bound substrates they used the local density function (LDF) theory, an ab initio quantum mechanical method developed in physics and recently applied to chemical problems. For the calculations the crystal structure of the *E. coli* DHFR.NADP.folate complex (69) was used, and modified to convert NADP to NADPH. Hydrogen atoms in standard geometries were added, and a hydration shell (5 Å around the protein, and 8 Å around the ligand atoms, with crystallographic waters included) was constructed. The structure was then subjected to energy minimization with constraints on the non-hydrogen atoms. In addition, the hydration shell was equilibrated using a short molecular dynamics simulation.

To study the effect of the electric field of the protein on the electron density distribution of folate, the electron density of the whole folate molecule in the hydrated enzyme complex was calculated (72). A second calculation was then carried out to determine electron density in the unbound ligand, with the molecule in the same conformation as in the complex. A difference electron density map was then calculated and analyzed. The result indicated that the enzyme induces significant polarization of the charge distribution in bound folate. The net change in charge for the atoms of the pteridine ring is $+0.61$, while there is a corresponding equal increase in negative charge for the atoms of the glutamate moiety, and no change in charge on the p-aminobenzoyl moiety. Within the pteridine ring it was found that the region around the C7—N8 bond (the bond to be reduced) became more positive. Moreover, electron density in the plane of the ring (at the level of the σ bond) was increased, whereas the π density above and below the plane of the ring was decreased.

These changes in the electron distribution at the C7—N8 bond are considered by Bajoreth et al. to be "consistent with the result of the forward enzymatic reaction (reduction of the C7—N8 double bond to a single bond)". The C6—N5 bond showed a significantly smaller region of positive density, with almost no effect on C6, while the electronic perturbations in the region of the N1—C2 and C4a—C8a double bonds were quite small.

An additional effect of the polarization of the bound folate was found to be an increase in the binding energy. Calculations indicated that the polarized folate binds about 4 kcal/mol more tightly to the enzyme than the unpolarized substrate. This result is consistent with the fact that the substrate in the transition state binds more tightly to the enzyme than in the ground state, but the increase in binding appears to be rather small (an increase by a factor of 966).

When similar calculations were applied to a model of the ternary complex of $E.$ $coli$ DHFR.NADP.H$_2$folate, similar but smaller polarization of the ligand was again predicted. A net change in charge of $+0.47$ was calculated for the pteridine ring, a change of $+0.05$ on the p-aminobenzoyl moiety, and a change of -0.53 on the glutamate moiety. When charge redistribution within the pteridine ring was examined, it was found to be greatest at C6 and N5, with increased electron density at the σ level, in the ring plane, with significant depletion of π electron density, above and below the plane of the ring. Other double bonds were less affected. This selective redistribution of electron density around the bond to be enzymically reduced is analogous to the results of the calculations for the folate complex.

When the electrostatic potential gradient for the whole complex was calculated (74), it was found that according to the calculations, despite a net charge of -10 on $E.$ $coli$ DHFR, there is a region of positive charge at the entrance to the active site pocket, close to the glutamate moiety of the bound substrate. This positive charge is primarily due to three basic side chains: Lys[32], Arg[52], and Arg[57]. If the calculation of the polarization of bound folate is performed with these three residues uncharged, the change in charge on the pteridine decreases from $+0.61$ to $+0.39$, indicating the importance of these residues for the polarization. There are also a number of enzyme groups in the vicinity of the pteridine ring that have partial negative charges: the carboxyl of Asp[27] (considered in the calculations to be protonated); the carbonyl oxygen atoms of Ile[5], Asp[27], and Ile[94];

and the hydroxyls of Tyr[100] and Thr[113]. If the calculation is performed with these groups uncharged as well as the basic residues, the change in charge on the pteridine ring drops still further to $+0.24$.

How valid are these conclusions? Although the LDF theory has previously been applied to small molecules, this is the first time it has been applied to a protein, so that there are no similar cases with which to compare the results. Furthermore, at the present time there are no obvious, alternative, independent methods available by which the conclusions regarding the selective redistribution of electron density around the bond to be reduced can be checked. If the positive charge field at the glutamate region of the pocket is indeed very important for substrate polarization, and if this in turn is the mechanism for the selective redistribution of electron density in the pteridine ring, then abolition of some, or all, of the positive charge by directed mutagenesis should greatly decrease the rate of the chemical transformation, but this has not yet been tested. One important simplification in the treatment is that no calculation was made of the effects of repolarization of the protein that would result from polarization of the ligand. This would certainly result in a decrease in the calculated polarization of the ligand by an unknown amount. It is also disturbing that in the calculation the active site carboxyl group (in this case the side chain carboxyl of Asp[27]) is protonated. This assumption is very improbable, at least at physiological pH, and the effect of this on the calculations is likely to be considerable.

If, nevertheless, these conclusions are valid, do they apply to eukaryotic DHFRs? The net charges on DHFRs from different sources are very different, ranging from -10 for the *E. coli* enzyme to $+3$ for porcine DHFR. In general the charge on bacterial DHFRs is more negative than that on the vertebrate DHFRs (Table XIII). Bajoreth et al. (74) point out that Arg[36], Lys[68], and Arg[70] in chicken DHFR form a very similar arrangement of positively charged residues around the entrance to the active site to that described above for *E. coli* DHFR. These residues are conserved in vertebrate DHFRs, and probably (as judged by the tentative sequence alignment in Fig. 1) in the yeast and *P. carinii* DHFR sequences also. Consequently, this facet of charge distribution on DHFR may be a conserved feature. However, there is no information as to how different the electrostatic charge distribution over the eukaryotic DHFRs is from that of *E. coli* DHFR, whether the bound substrate

TABLE XIII
Net Charge of Eukaryoyic DHFRs

Species	Net Charge
P. carinii	+11
S. cerevisiae	+5
C. neoformans	+3
Chicken	+1
Bovine	−1
Murine	−2
Porcine	+3
Human	0
Chinese hamster	+2

is still polarized despite the large differences in net charge, or whether LDF calculations for these DHFRs predict that the bond to be reduced in the bound substrate undergoes the same kind of electron density redistribution that is calculated for substrates bound to the bacterial enzyme.

B. PROTON DONATION TO THE SUBSTRATE

1. Does the Enzyme Preprotonate the Substrate?

The idea that DHFR facilitates the transfer of the hydride ion from NADPH to C6 or C7 of the substrate by first protonating the adjacent nitrogen dates back to a symposium presentation by Huennekens and Scrimgeour (75). That paper argued for preprotonation by drawing an analogy between reduction of NAD and reduction of folate and dihydrofolate. The false assumption was made that for N8 of folate and for N5 of H_2folate $pK_a \approx 9$–10, so that these nitrogen atoms would be protonated at neutral pH, with consequent depletion of electron density at C7 in folate and at C6 in dihydrofolate. The analogy with the pyridine nucleotide system was based primarily on the fact that nucleophiles like CN^-, thiols, and HSO_3^-, can add to C4 of NAD and to folate or dihydrofolate. However, very different conditions of pH are needed for these additions to the nicotinamide ring and to the pteridine ring. In the case of folate and dihydrofolate additions readily occur at pH near neutral without the intervention of enzymes (76, 77). Since Poe (78, 79) reported a pK_a of 2.35 for

N1 of folate (and no detectable ionization of N8), and pK_a of 3.84 for N5 of H_2folate, C7 in folate and C6 in H_2folate are electrophilic without protonation of the adjacent nitrogen. In fact, although we were able to confirm the pK_a reported by Poe for folate (80), careful redetermination of the pK_a values for H_2folate (81) showed that the pK_a for N5 of H_2folate is 2.56, making this point even more obvious.

The electrophilicity of C7 in folate and C6 in H_2folate, even when the pteridine ring bears no net charge, is due to the well-known polarization of the C$=$N double bond. This polarization is illustrated by ab initio calculations of atomic charges on neutral 6-methyl-7,8-dihydropterin (82), Values of $+0.32$ and $+0.26$ were obtained for C6 of the keto tautomer by alternate methods, while the charge on N5 was estimated as -0.60 or -0.47. Consequently, preprotonation of the substrate is not a prerequisite for hydride addition during the reaction catalyzed by DHFR. Furthermore, ab initio calculations indicate that protonation of the adjacent nitrogen depletes the electron density at C7 of folate or C6 of H_2folate by only a modest amount, increasing the positive charge by 0.12 or 0.17, respectively (83).

In view of these chemical considerations it seems possible that hydride could be added to bound substrate by DHFR without preprotonation of the substrate. In fact, substrates in binary complex with DHFR are not protonated. Although crystallography of the human DHFR.5-deazafolate complex exhibits evidence for protonation of N8 of the bound inhibitor, analysis of crystal diffraction data for the corresponding folate complex does not indicate protonation of the substrate (18). Moreover, when ^{15}N NMR was used to examine the protonation state of [5-^{15}N]H_2folate or [5-^{15}N]H_2biopterin complexed with bovine DHFR, the chemical shift data showed unequivocally that there is no detectable protonation of N5 of the substrate in the binary complex (84). This conclusion was further supported by ^{13}C NMR with [6-^{13}C]-labeled substrate complexes. The same conclusion was reached from Raman difference spectra (85).

However, protonation of substrate in ternary complexes must also be considered. When the human DHFR.NADPH complex was rapidly mixed with a less than stoichiometric concentration of dihydrobiopterin and spectral changes monitored by stopped flow spectrophotometry, there were no changes indicative of protonation of the dihydrobiopterin (22). Since the rate of conversion of bound dihydrobiopterin to tetrahydrobiopterin is relatively slow, it was expected that rapid formation of preprotonated dihydrobiopterin would be detectable before acceptance of a hydride ion. However, a different conclusion was reached by Chen et al. (85) who used Raman difference spectroscopy to investigate protonation of dihydrofolate in the ternary complex with *E. coli* DHFR and NADP. Three prominent bands due to bound dihydrofolate were observed at 1608, 1650, and 1675 cm^{-1}. The 1608-cm^{-1} band is pH insensitive and was assigned to the benzoyl group. The other two bands are assigned to the N5=C6 stretch of the pteridine ring of dihydrofolate. As the pH of the solution is decreased from 9.2 to 5.4 the intensity of the 1650-cm^{-1} band decreases while that of the 1675-cm^{-1} band increases. The bands are therefore assigned to the unprotonated and protonated forms of bound dihydrofolate, respectively. Confirmation of this assignment rests on; (a) a shift of the 1675-cm^{-1} band to 1661 cm^{-1} in D_2O, with the 1650-cm^{-1} band unaffected; (b) both bands shifted to lower frequency by the expected amount for bound [6-^{13}C]dihydrofolate; (c) a band at 1635 cm^{-1} for unbound dihydrobiopterin at pH 5 (i.e., with N5 unprotonated) migrating to 1671 cm^{-1} at pH 1.6 (i.e., with N5 protonated); (d) a shift of both bands to lower frequency for [5-^{15}N]dihydrobiopterin.

The change in intensity of the 1650 and 1675-cm^{-1} bands of bound dihydrofolate with pH gives a pK_a of 6.5 for protonation of N5 of dihydrofolate in the ternary complex.

The 1675-cm^{-1} band is not observed at pH 6.1 for dihydrofolate in the binary complex with *E. coli* DHFR, or in the ternary complex with NADPH$_4$ (an inactive analog) also present, or in the ternary complex with NADP and the D27S variant of *E. coli* DHFR. It is suggested that the enzyme, especially Asp^{27} (which corresponds to Glu^{30} in eukaryotic DHFR), and the bound cofactor together create a local electrostatic field that raises the pK_a of N5 of bound dihydrofolate from 2.6 in the absence of enzyme to 6.5 in the ternary complex. It would be desirable for this finding to be confirmed by other means, and to determine whether similar results are obtained with vertebrate DHFR, and with other complexes.

2. Is the Enzyme the Immediate Proton Donor?

A considerable literature exists suggesting that the active site acidic side chain (Glu^{30} in vertebrate DHFR, and Asp^{27} in *E. coli*

DHFR) is involved in proton donation to the substrate. However, there are severe problems with this view, and it will be critically examined.

Stone and Morrison (86, 87) reported that k_{cat} for *E. coli* DHFR shows pH dependence, with velocity dropping off quite steeply as the pH is decreased below pH 8. This result was later confirmed by Howell et al. (88) and by our own laboratory (67), except that the results of the latter two groups indicated that k_{cat} also *decreases* as the pH decreases below 7. To interpret their results Stone and Morrison made a number of proposals: (a) that the pH dependence of k_{cat} is described by a single protonation, with pK_a of 8.4; (b) that this protonation occurs on the enzyme; (c) that the group protonated is Asp[27]; (d) that this protonated carboxyl group is the immediate proton donor to N5 of H_2folate; and (e) that this protonation precedes and facilitates hydride transfer. Evidence for and against these proposals will now be discussed.

The first modification to this proposed scheme was provided by the work of Fierke et al. (66) who showed that the increase in k_{cat} as the pH decreases from 10 is due to an increase in the rate constant of the chemical transformation step (k_{chem}). The apparent pK_a of 8.4 does not represent a true pK_a, but is the pH at which k_{chem} becomes equal to k_{off} for product release. The pH dependence of k_{chem}, over the range pH 5–9, fits quite well to a single protonation with a pK_a of 6.5. This pK_a was again attributed by Fierke et al. (66) to the carboxyl group of Asp[27], and it was again assumed that this carboxyl group donated its proton to substrate.

Seeming support for this view came from the report of Howell et al. (88) on two variants of *E. coli* DHFR with substitution of Asn or Ser for Asp[27]. Crystallography of the methotrexate complexes of these variants revealed that the structures of the wild-type enzyme and the two variants show no difference except in the immediate vicinity of the substituted side chains. In the complex of the Asn[27] variant the Asn side chain makes strong hydrogen bonds with the 2-amino group and N3 of MTX (Scheme 5), just as in the case of the Asp[27] side chain in the complex of wild-type enzyme, but the different hydrogen bonding of the amido group of Asn[27] to a water molecule (Wat[403]) causes movement of the latter by 0.9 Å, as well as movement of another fixed water to which Wat[403] is in turn bonded. In the Ser[27] variant there is a greater structural difference in the immediate vicinity of the Ser side chain, which is not hydrogen bonded to MTX. A new, fixed water molecule occupies a position close to that occupied by one of the Asp oxygen atoms, and is hydrogen bonded both to N1 of MTX and to the hydroxyl of Ser^{27} (Scheme 5). The MTX is protonated at N1 in its complexes with bacterial and vertebrate DHFRs throughout the accessible pH range (23). Although the pteridine ring of MTX in its complexes with the Asn^{27}, and Ser^{27} variants occupies a similar position to that in the complex with wt enzyme, it remains unprotonated throughout the accessible pH range (89), so that there is no ionic interaction between MTX and the enzyme.

The Michaelis constant for H_2folate and k_{cat} are both changed by the side chain substitutions. At pH 7.0 and 30°C the values for k_{cat} are wild type, 30; Asn^{27}, 0.1; Ser^{27}, 0.44 s^{-1}. The K_m values are 1.2, 44, and 140 μM, respectively. The most striking change is in the effect of pH on k_{cat}, which continuously increases as the pH is decreased, and approaches that of wt at pH 5. The E30Q variant of human DHFR has similar behavior to the D27N variant of *E. coli* DHFR. The Michaelis constant for H_2folate is greatly increased (100 μM) and k_{cat} is greatly decreased (0.3 s^{-1}) at pH 7.65, but increases to 30 s^{-1} at pH 5 (51). Howell et al. (88) interpreted their results to mean that neutral H_2folate bound to wild-type enzyme is readily protonated through the mediation of Asp^{27} before hydride transfer can occur, but in the case of the variant DHFRs only binding of already protonated H_2folate is productive. However, on this view, k_{cat} should increase 10-fold for each decrease in pH by one unit, until the pH is near 2.56, the pK_a for unbound H_2folate, so that plots of log k_{cat} versus pH should be linear with a slope of -1. In fact, such plots for the variants approximate poorly to a straight line, and the slope of the best fit lines is -0.6 to -0.7.

There are several possible explanations for this discrepancy. One is that the expected pH dependence really refers to k_{chem}, and this may not be equal to k_{cat} over all the pH range. However, data provided by Howell et al. indicate that $k_{chem} \approx k_{cat}$ for the Ser^{27} variant at pH $\geq$ 5, and over most of this pH range for the Asn^{27} variant. Similar plots of log k_{cat} versus pH with slopes of 0.5–0.65 were reported for variants of *E. coli* with substitutions of Ile, Gly, or Asn for Leu^{54} (90). For these variants chemical transformation is definitely rate limiting. The plots of k_{cat} versus pH cannot therefore be

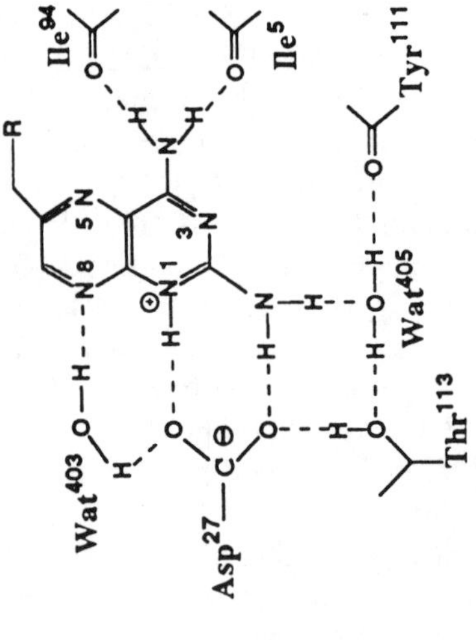

Asp27.MTX

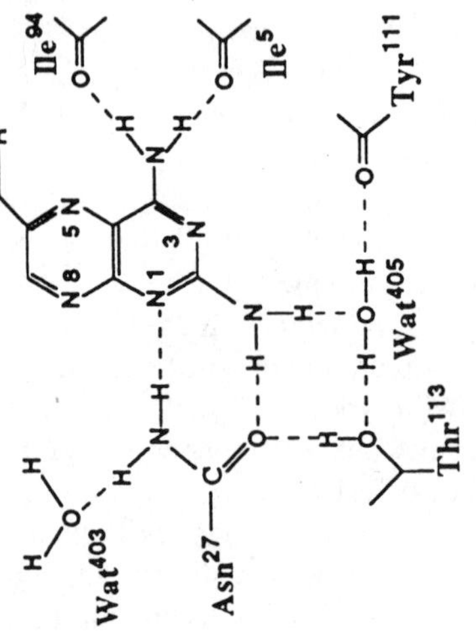

Asn27.MTX

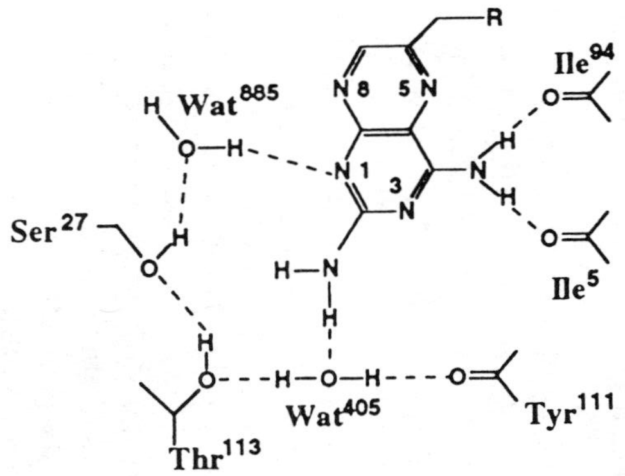

Ser²⁷·MTX

Scheme 5. Hydrogen bonding and ionic interaction between MTX and wild-type (Asp²⁷) *E. coli* DHFR, and hydrogen bonding between MTX and variants (Asn²⁷ and Ser²⁷) of *E. coli* DHFR (88).

explained adequately in this fashion. Use of the correct pK_a for N5 of H_2folate rather than the erroneous value in the older literature does not improve the fit of the data to the theory either (81). Howell et al. suggest that the discrepancy is due to a pH-dependent equilibrium between apoenzyme conformers, but this should not have affected results if the enzyme was equilibrated with one of the substrates before the reaction commenced, since this would pull all of the enzyme into the tight-binding conformer. It seems likely that the effect of pH on the reaction is therefore a good deal more complicated than the model of Morrison and Stone proposes. If the effect of pH on k_{chem} for wt DHFR corresponds to protonation of N5 of H_2folate in the ternary substrate complex as proposed by Chen et al. (85), it would not be unexpected that pH dependence would be more complex than that of a single carboxyl group.

The crystal structure of the folate complexes with DHFR made it quite clear that neither Glu[30] in vertebrate DHFRs (Fig. 2), nor Asp[27] in bacterial DHFRs is positioned so that the side chain carboxyl group can donate a proton to N8 of folate, or to N5 of H_2folate. This result is conclusive evidence against the view that the protonated enzyme is the immediate proton donor to the substrate. The "enzyme as proton donor" hypothesis was then changed to propose that the protonated carboxyl group facilitates the conversion of the bound substrate from the imino-keto to the enol tautomer and that the enol group is the immediate donor of the proton to N5 of H_2folate (82, 87, 91). However, when the crystal structures of the chicken DHFR.NADP.biopterin complex (15) and other substrate complexes were examined, it became clear that the geometry is very unfavorable for direct transfer of a proton from O4 of the pterin to N5. Also our ^{15}N NMR results established that in binary complexes involving human DHFR and folate or H_2folate labeled in the 2-amino group and at N3 with ^{15}N the proton remains on N3 and exchanges very slowly, in contrast to the rapid exchange from the unbound ligands (92). The same is true for the labeled E.folate.NADP complex. This is the case even in the high pH range in which this proton ionizes readily from free folate, and it is also true at pH 5 where the rate of hydride transfer is near its maximum. Cheung et al. (93) also concluded from ^{13}C NMR of [4, 6, 8a-^{13}C]- and [2, 4a, 7, 9-^{13}C]folate complexed to *L. casei* DHFR that folate bound in the productive orientation is present in the imino-keto form at pH 6.5. There is therefore no evidence to suggest that in the enzyme complex there is any change from the normal imino-keto tautomer. In fact the substrate appears to be locked by the enzyme even more tightly into its imino–keto structure by hydrogen bonding with the carboxyl group of Glu[30].

pH dependence of k_{chem} seems to be best understood as follows. As the pH changes, there are changes in both the net charge on the enzyme and the electrostatic potential gradient over its surface and in the active site. This in turn may change the polarization of the bound substrate, with consequent redistribution of electrons in the bond to be reduced (Section V.A.3), which affects the rate of the bond reduction. It may also change the equilibrium for proton transfer to substrate in the ternary complex. But it is not necessarily a direct measure of the protonation of the active site carboxyl group. Such an interpretation is consistent with the fact that when folate or H_2folate isotopically labeled at C2, N3, or the 2-amino group is bound to human DHFR, NMR does not detect protonation of the nearby carboxyl group of Glu[30] over the pH range 5–7 (92).

3. Bound Water as the Immediate Proton Donor to Substrate

McTigue et al. (15) noted that in the ternary complex of chicken DHFR with NADP and biopterin the pterin ring of bound biopterin is linked to bulk solvent by two series of ordered water molecules. One solvent channel has as its terminus Wat[230] (Fig. 4), a molecule that is hydrogen bonded to the side chain carboxylate of Glu[30], the indole N of Trp[24], and O4 of biopterin. A fixed water is in the same position in all the known DHFR structures. The solvent channel from Wat[230] to bulk solvent is bordered by residues 22–30. This is the channel that has been the focus of attention in previous discussions of substrate protonation.

However, the other water channel has as its terminus Wat[756] (Fig. 4). This water molecule is not seen in the other crystal structures, although its transitory entry for substrate protonation was postulated by Bystroff et al. (69) in their mechanistic scheme based on the crystal structure of the E. coli DHFR.NADP.folate complex. This water molecule makes strong hydrogen bonds with both O4 and N5 of the pterin ring and is the obvious (and necessary) proton donor to N5. In the biopterin ternary complex, Wat[756] is also hydrogen bonded with the two side chain hydroxyls, but its other hydrogen bonds suffice to make its position energetically possible in the folate or dihydrofolate complex. The solvent channel to Wat[756] is bordered by the side chain of Tyr[31] and residues 59–61. It is perhaps significant that Tyr[31] in chicken DHFR, and the corresponding Phe[31] in human DHFR, adopt alternate distinct conformations in the crystal structures, perhaps indicating that they can move to allow entry and exit of the water molecule.

There seems little doubt that Wat[756] in the chicken DHFR complex, and a corresponding water that is admitted into the other complexes during the reduction cycle, together with the water channel that connects them to bulk solvent fulfills all the requirements for the system for donation of a proton to N5 of dihydropterin substrates. The other fixed water molecule (Wat[230] in the chicken DHFR structure) is not near enough to be the proton source, and an enolic group on the pterin will not do as a means of transmitting the proton between Wat[230] and N5 of the substrate.

C. THE ISOMERIZATION COMPONENT OF THE CHEMICAL TRANSFORMATION

During the first turnover of enzyme-bound substrates to enzyme-bound products the rate of the reaction is quite independent of product release from the enzyme, although the rate of product release has an important, and indeed often dominant, role in determining the rate of subsequent product formation. Consequently, the rate of the first turnover should depend only on the rate of hydride and proton transfer to the pterin from the dinucleotide. Since proton transfer is expected to be rapid, the rate of the first turnover should equal the rate of hydride turnover and be subject to the same isotope effect on substituting NADPD for NADPH. However, this is not always the case.

When the isotope effect for the first turnover was determined for bacterial DHFRs at pH 7.65, it was in the range 3.0–3.2, which is close to the value of the isotope effect on k_{cat} at pH 10, conditions in which hydride transfer appears to be rate limiting (67). In contrast to this result the isotope effect on the rate constant for the first turnover for vertebrate DHFRs is only 2.1–2.3. Several possible explanations for this were considered. It was demonstrated that the rate of the first turnover is not limited by the rate of substrate binding, which is very fast. The intrinsic isotope effect was demonstrated to be greater than 2.1–2.3, not only from results with the bacterial DHFRs, but also from the fact that the isotope effect on k_{cat} for chicken DHFR is 3 at pH 10. Consequently, it was concluded that an isomerization of the ternary *substrate* complex precedes the chemical transformation step. An isomerization of the ternary product complex is a less likely explanation of the low isotope effect, although both isomerizations might occur. Although there is no evidence for a similar isomerization influencing the rate of the first turnover for bacterial DHFRs at pH 7.65, at pH 5.0 the isotope effects are only 2.1 for *L. casei* DHFR and 1.4 for *E. coli* DHFR, so that an isomerization appears to occur in the mechanism of the bacterial enzymes also, but is silent at neutral pH because it is much faster than hydride and proton transfer at such a pH.

Similar evidence has been presented previously by others for such isomerizations of the ternary substrate complexes of some of the classical dehydrogenases. Very low isotope effects were reported for the first turnover for lactate dehydrogenase (94) and for phosphoglycerate dehydrogenase (95).

The most obvious interpretation of this isomerization is that it involves small structural movements concerned either with entry into the transition state, or with proton donation, or both. As discussed in Section V.A.2, Kraut and co-workers proposed that in entering the transition state both the pterin ring and the nicotinamide ring rotate slightly in order to optimize the distance of C4 of the nicotinamide ring from C6 of the H_2folate pteridine ring for hydride transfer, the optimum distance being approximately 2.6 Å. If the theory is correct, such movement would precede hydride and proton transfer, so that this might constitute the isomerization.

It is argued in Section V.B.3, following Bystroff, McTigue, Kraut et al. (15, 69), that a water molecule is admitted during each catalytic cycle to the active site where it hydrogen bonds with N5 and O4 of the bound pterin substrate, taking up a position similar to that occupied by Wat[756] in the chicken DHFR NADP.biopterin complex, before donating a proton to N5 of the substrate in the transition state. The admission and expulsion of this water is an alternative explanation of the isomerization process integral to the chemical transformation.

There is currently insufficient information to determine which, if either, of these possibilities is the correct interpretation of the isomerization reaction. The rate of the isomerization is perhaps more consistent with the second explanation, but this is by no means certain.

D. EFFECTS OF MUTATIONS ON THE RATE OF THE CHEMICAL TRANSFORMATION

The effects on k_{chem} of mutations causing amino acid substitutions at the active site of human DHFR have already been summarized in Table XI. The position in the active site of the residues substituted has already been discussed in an earlier section (III.C). It may be seen from Table XI that quite conservative replacements of any of several active site residues cause a large decrease in k_{chem}, and in fact few, if any, replacements can be made without a large decrease in k_{chem}. Similarly, Wagner et al. (96) showed that the F31L variant of mouse DHFR has a k_{chem} 56% that of the wt. It is reasonable to conclude therefore, that the whole architecture of the active site influences the process of proton donation to substrate, or hydride addition, or isomerization of the ternary substrate complex, or two or more of these processes. Although it has been argued in preceding sections that the electrostatic potential gradient in the active site is critical for the chemical transformation, many of the substitutions that greatly decrease k_{chem} involve replacement of one nonpolar residue by another. Conceivably such substitutions do affect the electrostatic potential gradient by perturbing the number and position of fixed water molecules.

VI. Alternate Substrates

Although no other substrate is reduced by eukaryotic DHFRs as rapidly as H_2folate and its polyglutamate forms, the unreduced parent compound, folate, is reduced at a sufficient rate in the intestinal mucosa that physiological doses of folate are reduced and converted to methyl and formyl derivatives of H_4folate by the time the absorbed folate reaches the serosal side (97). At high lumen concentrations of folate, however, most of the transported folate is unmodified and is carried in the blood to the liver where it is rapidly reduced. This is the metabolic explanation of the fact that although there is virtually no unmodified folate in tissues, this parent compound can meet the nutritional needs of higher animals for folates.

Although H_2biopterin does not lie on the biosynthetic pathway that provides H_4biopterin to act as cofactor for the hydroxylation of phenylalanine, tyrosine, and tryptophan (98), some H_2biopterin is probably formed by adventitious oxidation of H_4biopterin, or of 6-lactoylH_4pterin, or by isomerization of quininoid H_2biopterin, and for this loss to be recovered, DHFR must reduce the H_2biopterin. Consequently, the reduction of these and related substrates by vertebrate DHFRs is of interest.

A. FOLATE AS SUBSTRATE

Steady-state kinetic parameters for folate reduction have been determined for several vertebrate DHFRs (Table XIV), but the values have been determined at different pH values, or a pH independent value has been calculated, so that comparisons are difficult, since k_{cat} is greatly affected by pH.

It may be seen that K_m for folate is much higher than for H_2folate, and k_{cat} is very much lower. It is again difficult to make strict comparisons of the parameters for the two substrates, because of the different reaction conditions often used, but in our hands the human enzyme K_m is about 10 times greater for folate than for H_2folate, and k_{cat} 893 times lower for folate at pH 7.65.

TABLE XIV
Steady State Kinetic Parameters for Folate Reduction

| Enzyme Source | K_m (μM) | | k_{cat} (s^{-1}) | pH | Reference |
	Folate	NADPH			
Human	1.11	0.42	0.014, 0.37[a]	7.65	22
	0.43[a]		0.0013[a]		35
Chicken	6.7[b], 1.6[c]			4.0	63
Mouse	10		0.44	6.0	32
Rat	0.88		0.075	6.5	65

[a] pH independent value; applicable, for example, at pH 5.

[b] Citrate buffer.

[c] Acetate buffer.

It may be seen from Table VI that the although the binding of folate is weaker than binding of H_2folate, it is weaker only by a factor of 1.4–5 for various vertebrate DHFRs, and only partly accounts for the increase in K_m. When the contributions to the lower K_d are examined it can be seen (Table VII) that the rate constants for folate association with DHFR and its dinucleotide complexes are about one-half the corresponding k_{on} values for H_2folate association. In addition, the value of k_{off} for the binary complex is higher than that for the H_2folate binary complex. However, k_{off} for the folate-NADP ternary complex is much lower than for the H_2folate complex, providing most of the basis for the very low K_d (from k_{off}/k_{on}).

The value of k_{chem} for human DHFR at pH 7.65 is about 0.014 s^{-1}, which is 97,000 times lower than for H_2folate. The cause of this very low rate is uncertain. When NADPH was modeled into the crystal structure of the folate binary complex of human DHFR to obtain a proposed model of the transition state (18), C4 of the reduced nicotinamide ring of NADPH was found to be almost equidistant from C6 and C7 of the folate ring (2.6 and 2.4 Å, respectively). This model was supported by the subsequently published crystal structure of the chicken DHFR.NADP.biopterin complex (15). In this structure C4 of the nicotinamide ring of NADP is positioned 3.31 Å from C7 of biopterin, and 3.47 Å from C6 of biopterin. Similarly, in the crystal structure of the ternary E. coli DHFR.NADP.folate complex C4 of the nicotinamide ring of NADP is 3.18 Å from C7 and 3.30 Å from C6 of folate (69). These distances give no indication that hydride transfer to C6 would be more favorable than transfer to C7. In theoretical studies on the transfer of hydride from dihydropyridine to the methyleniminium cation by Wu and Houk (70), it was calculated that a syn configuration of the acceptor C—N bond with respect to the C4-N1 axis of the dihydropyridine ring favors hydride transfer. However, examination of the crystal structures of the ternary complexes already referred to does not appear to indicate more favorable geometry for transfer to C6 compared to transfer to C7. It must be emphasized that the structures examined are in the ground state, whereas hydride transfer occurs in the transition state, but it seems unlikely that the geometry changes greatly in attainment of the transition state. Nevertheless, hydride transfer is largely rate limiting in folate reduction since there is a relatively high isotope effect for k_{cat}. It is 3.0 at pH 7.65 and 1.8 at pH 5 for the human enzyme (22), and 3 at pH 6–7 for the mouse enzyme (32). One explanation for the poor rate of hydride transfer may be that the 7,8 double bond in folate is less polarized than the 5,6 double bond in H_2folate. This seems highly likely since no nucleophilic additions are known to occur across the 7,8 bond of folate or of other pterins like the additions occurring at the 5,6 bond of H_2folate. However, the effective pK_a of N8 of folate in the ternary substrate complex does not seem to be greatly affected by this, since the pK_a for the dependence of k_{cat} on pH is 6.09 (22), not much lower than that for dihydro substrates.

The rate constants for human DHFR given in Table VII, together with that for the chemical transformation, lead to the reaction scheme for human DHFR at pH 7.65 shown (Scheme 6). It is clear that this pathway has many branches that make significant contributions. Starting with the conversion of the ternary substrate complex to the ternary product complex, the low k_{chem} may be predicted to result in a significant fraction of the enzyme being sequestered as

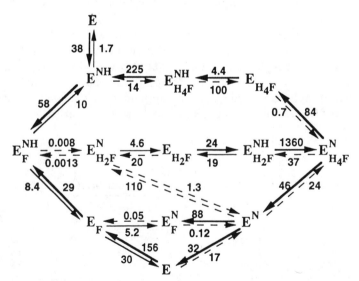

Scheme 6.   Kinetic scheme for reduction of folate by human DHFR at 20°C and pH 7.65. *Bold arrows* indicate major pathways, *broken arrows* indicate minor pathways (22). Units for rate constants and abbreviations as in Scheme 3.

the substrate ternary complex at steady state. In addition, however, there are a number of other aspects of the pathways that adversely affect the reduction of folate. The release of either product from the E.NADP.H$_2$folate complex is slow, so that there will be some accumulation of this complex at steady state also. Dissociation of NADP from this complex, the major pathway, permits rapid binding of NADPH, and a rapid cycle by the normal route for H$_2$folate reduction (upper loop in Scheme 6). However, a significant fraction of the E.NADP.H$_4$folate will dissociate H$_4$folate to give E.NADP, and some of this complex will also be formed from E.NADP.H$_2$folate. Although the binary NADP complex dissociates at a fairly high rate (32 s^{-1}), much of the binary complex will be converted to the E.NADP.folate dead-end complex, from which either ligand can dissociate only very slowly, so that a significant fraction of enzyme will accumulate in the form of this complex.

B. SUBSTRATES WITH SHORTER SIDE CHAINS

Relatively few other substrates have been explored with eukaryotic DHFRs, and the information available about the kinetics of their reduction is sparse. Some steady-state kinetic parameters are collected in Table XV. It may be seen that in all cases K_m is high, and where it is known, k_{cat} is low. The data presented in Table VII indicate that a very high k_{off} is mainly responsible for the weak binding of H$_2$biopterin. Thibault et al. (99) argue that the reduction of 8-methylpterin occurs because this compound is an analog of the protonated form of folate. The analog has a much lower K_m at pH 5

TABLE XV
Steady State Kinetic Parameters for Substrates with Short Side Chains

Ligand	Enzyme Source	K_m (μM)	k_{cat} (s^{-1})	pH	Reference
H$_2$biopterin	Human	21	9.0	5.0	35
	Rat	6.4	0.81	6.5	65
6-MethylH$_2$pterin	Rat	10.2	0.37	6.5	65
8-Methylpterin	Human	310		7.4	99
	Chicken	217		7.4	99
	Bovine	200		7.4	99

than at pH 7.4 for all three DHFRs, and at the pH optimum the rate of reduction is reported to be greater than for H_2folate reduction.

VII. Concluding Remarks

Although recent years have seen a great increase in the available information about eukaryotic DHFRs, especially about the mammalian enzymes, there remains a good deal to be learned. Valuable insights have been obtained to the mechanism by which the enzyme facilitates transfer of a proton and hydride ion to the substrate, but much, perhaps including the most important facts, about this process remains to be discovered. Details of how the precise structure of the enzyme is linked to the strength of ligand binding, and to the efficiency of the chemical transformation step are still understood very incompletely.

These and other aspects of this group of enzymes are of significance not only to those interested in enzyme structure–function relationships, but also to those who see therapeutic possibilities in variants of DHFR with properties changed in specific directions, for example, variants that retain high catalytic efficiency but are inhibited weakly by certain therapeutic drugs. Consequently, it is anticipated that there will continue to be interest in this area for some time to come.

Acknowledgments

I am indebted to Dr. J. R. Appleman for many helpful suggestions and critical comments regarding the text, as well as for his stimulating collaboration in work we have performed in this area. Our research in this area is supported in part by United States Public Health Service Research Grant RO1 CA 31922 and Cancer Core Grant P30 CA 21765, and by American Lebanese Syrian Associated Charities (ALSAC).

References

1. Edman, J. C., Edman, U., Cao, M., Lundgren, B., Kovacs, J. A., and Santi, D. V., *Proc. Natl. Acad. Sci. USA,* **86,** 8625–8629 (1989).
2. Fling, M. E., Kopf, J., and Richards, C. A., *Gene,* **63,** 165–174 (1988).

3. Barclay, B. J., Huang, T., Nagel, M. G., Misener, V. L., Game, J. C., and Wahl, G. M., *Gene,* **63,** 175–185 (1988).

4. Sirawaraporn, W., Cao, M., Santi, D. V., and Edman, J., Cloning, *J. Biol. Chem.,* **268,** 8888–8892 (1993).

5. Kumar, A. A., Blankenship, D. T., Kaufman, B. T., and Freisheim, J. H., *Biochemistry,* **19,** 667–678 (1980).

6. Lai, P.-H., Pan, Y.-C. E., Gleisner, J. M., Peterson, D. L., Williams, K. R., and Blakley, R. L., *Biochemistry,* **21,** 3284–3294 (1982).

7. Stone, D., Paterson, S. J., Raper, J. H., and Phillips, A. W., *J. Biol. Chem.,* **254,** 480–488 (1979).

8. Smith, S. L., Patrick, P., Stone, D., Phillips, A. W., and Burchall, J. J., *J. Biol. Chem.,* **254,** 11475–11484 (1979).

9. Masters, J. N. and Attardi, G., *Gene,* **21,** 59–63 (1983).

10. Melera, P. W., Davide, J. P., and Oen, H., *J. Biol. Chem.,* **263,** 1978–1990 (1988).

11. Blakley, R. L., Dihydrofolate reductase, in *Folates and Pterins, Vol. 1, Chemistry and Biochemistry of Folates,* Blakley, R. L. and Benkovic, S. J., Eds., Wiley, New York, 1984, pp. 191–253.

12. Matthews, D. A., Bolin, J. T., Burridge, J. M., Filman, D. J., Volz, K. W., Kaufman, B. T., Beddell, C. R., Champness, J. N., Stammers, D. K., and Kraut, J., *J. Biol. Chem.,* **260,** 381–391 (1985).

13. Volz, K. W., Matthews, D. A., Alden, R. A., Freer, S. T., Hansch, C., Kaufman, B. T., and Kraut, J., *J. Biol. Chem.,* **257,** 2528–2536 (1982).

14. Matthews, D. A., Bolin, J. T., Burridge, J. M., Filman, D. J., Volz, K. W., and Kraut, J., *J. Biol. Chem.,* **260,** 392–399 (1985).

15. McTigue, M. A., Davies, J. F., II, Kaufman, B. T., and Kraut, J., *Biochemistry,* **31,** 7264–7273 (1992).

16. Stammers, D. K., Champness, J. N., Bedell, C. R., Dann, J. G., Eliopoulos, E., Geddes, A. J., Ogg, D., and North, A. C. T., *FEBS Lett.,* **218,** 178–184 (1987).

17. Oefner, C., D'Arcy, A., and Winkler, F. K., *Eur. J. Biochem.,* **174,** 377–385 (1988).

18. Davies, J. F., Delcamp, T. J., Prendergast, N. J., Ashford, V. A., Freisheim, J. H., and Kraut, J., *Biochemistry,* **29,** 9467–9479 (1990).

19. Cody, V., Luft, J. R., Ciszak, E., Kalman, T. I., and Freisheim, J. H., *Anti-Cancer Drug Design,* **7,** 483–491 (1992).

20. Tsay, J.-T., Appleman, J. R., Beard, W. A., Prendergast, N. J., Delcamp, T. J., Freisheim, J. H., and Blakley, R. L., *Biochemistry,* **29,** 6428–6436 (1990).

21. Subramanian, S. and Kaufman, B. T., *Proc. Natl. Acad. Sci. USA,* **75,** 3201–3205 (1978).

22. Appleman, J. R. and Blakley, R. L., unpublished results.

23. Cocco, L., Roth, B., Temple, C., Jr., Montgomery, J. A., London, R. E., and Blakley, R. L., *Arch. Biochem. Biophys.,* **226,** 567–577 (1981).

EUKARYOTIC DIHYDROFOLATE REDUCTASE 99

24. Margosiak, S. A., Appleman, J. R., Santi, D. V., and Blakley, R. L., *Arch. Biochem. Biophys., 305,* 499–508 (1993).

25. Matthews, D. A., Alden, R. A., Bolin, J. T., Filman, D. J., Freer, S. T., Hamlin, R., Hol, W. G. J., Kisliuk, R. L., Pastore, E. J., Plante, L. T., Xuong, N.-H., and Kraut, J., *J. Biol. Chem., 253,* 6946–6954 (1978).

26. Rossman, M. G., Liljas, A., Bränden, C.-I., and Banasak, L. J., Evolutionary and structural relationships among dehydrogenases in *The Enzymes,* vol. 9, Part A, Boyer, P. D., Ed., Academic, New York, 1975, pp. 61–102.

27. Kraut, J. and Matthews, D. A., Dihydrofolate reductase, in *Biological Macromolecules and Assemblies,* Vol. 3, Jurnak, F. A. and McPherson, A., Eds., Wiley, New York, 1987, pp. 1–72.

27a. Kaufman, B. T., Kumar, A. A., Blankenship, D. T., and Freisheim J. H. *J. Biol. Chem., 255,* 6542–6545 (1980).

27b. McTigue, M. A., and Kraut, J., unpublished results.

28. Stockman, B. J., Nirmala, N. R., Wagner, G., Delcamp, T. J., DeYarman, M. T., and Freisheim, J. H., *Biochemistry, 31,* 218–229 (1992).

29. Falzone, C. J., Wright, P. E., Benkovic, S. J., *Biochemistry, 30,* 2184–2191 (1991).

30. Appleman, J. R., Howell, E. E., Kraut, J., Kühl, M., and Blakley, R. L., *J. Biol. Chem., 263,* 9187–9198 (1988).

31. Appleman, J. R., Beard, W. A., Delcamp, T. J., Prendergast, N. J., Freisheim, J. H., and Blakley, R. L., *J. Biol. Chem., 265,* 2740–2748 (1990).

32. Thillet, J., Adams, J. A., and Benkovic, S. J., *Biochemistry, 29,* 5195–5202 (1990).

33. Prendergast, N. J., Appleman, J. R., Delcamp, D. J., Blakley, R. L., and Freisheim, J. H., *Biochemistry, 28,* 4645–4650 (1989).

34. Stone, S. R. and Morrison, J. F., *Biochim. Biophys. Acta, 869,* 275–285 (1986).

35. Williams, E. A., and Morrison, J. F., *Biochemistry, 31,* 6801–6811 (1992).

36. Birdsall, B., Burgen, A. S. V., de Miranda, J. R., and Roberts, G. C. K., *Biochemistry, 17,* 2102–2110 (1978).

37. Birdsall, B., Burgen, A. S. V., and Roberts, G. C. K., *Biochemistry, 19,* 3732–3737 (1980).

38. Baccanari, D. P., Daluge, S., and King, R. W., *Biochemistry, 21,* 5068–5075 (1982).

39. Bystroff, C. and Kraut, J., *Biochemistry, 30,* 2227–2239 (1991).

40. Penner, M. H. and Frieden, C., *J. Biol. Chem., 260,* 5366–5369 (1985).

41. Appleman, J. R., Beard, W. A., Delcamp, T. J., Prendergast, N. J., Freisheim, J. H., and Blakley, R. L., *J. Biol. Chem., 264,* 2625–2633 (1989).

42. Beard, W. A., Appleman, J. R., Huang, S., Delcamp, T. J., and Blakley, R. L., *Biochemistry, 30,* 1432–1440 (1991).

43. Appleman, J. R., Prendergast, N., Delcamp, T. J., Freisheim, J. H., and Blakley, R. L., *J. Biol. Chem., 263,* 10304–10313 (1988).

44. Thillet, J., Absil, J., Stone, S. R., and Pictet, R., *J. Biol. Chem.*, **263**, 12500–12508 (1988).

45. Srimatkandada, S., Schweitzer, B. I., Moronson, B. A., Dube, S., and Bertino, J. R., *J. Biol. Chem.*, **264**, 3524–3528 (1989).

46. Drake, J. C., Allegra, C. J., Baram, J., Kaufman, B. T., and Chabner, B. A., *Biochem. Pharmacol.*, **36**, 2416–2418 (1987).

47. Morrison, J. F., Dihydrofolate reductase, in *A Study of Enzymes*, Vol. II, Kuby, S. A., Ed., CRC Press: Boca Raton, FL, 1990 pp. 193–226.

48. Morrison, J. F., *Trends Biochem. Sci.*, **7**, 102–105 (1982).

49. Taira, K., Fierke, C. A., Chen, J.-T., Johnson, K. A., and Benkovic, S. J., *Trends Biochem. Sci.*, **12**, 275–278 (1987).

50. Chunduru, S. K., Appleman, J. R., and Blakley, R. L., *J. Biol. Chem.*, **269**, 9547–9555 (1994).

51. Delcamp, T. J., Freisheim, J. H., Appleman, J. R., and Blakley, R. L., unpublished results.

52. Appleman, J. R., Tsay, J.-T., Freisheim, J. H., and Blakley, R. L., *Biochemistry*, **31**, 3709–3715 (1992).

53. Chunduru, S. K., Appleman, J. R., and Blakley, R. L., *Adv. Exp. Med. Biol.*, **338**, 507–510 (1993).

54. Blakley, R. L., Appleman, J. R., Chunduru, S. K., Nakano, T., Lewis, W. S., and Harris, S. E., *Adv. Exp. Biol. Med.*, **338**, 473–479 (1993).

55. Cody, V., Wojtczak, A., Kalman, T. I., Freisheim, J. H., and Blakley, R. L., *Adv. Exp. Biol. Med.*, **338**, 481–486 (1993).

56. Nakano, T., Appleman, J. R., and Blakley, R. L., *Adv. Exp. Biol. Med.*, **338**, 503–505 (1993).

57. Nakano, T., Spencer, H. T., Appleman, J. R., and Blakley, R. L., *Biochemistry*, **33**, 9945–9952 (1994).

58. Lewis, W. S., Cody, V., Galitsky, N., Luft, J. R., Pangborn, W., Chunduru, S. K., Spencer, H. T., Appleman, J. R., and Blakley, R. L., *J. Biol. Chem.*, **270**, 5057–5064 (1995).

59. Thompson, P. D., and Freisheim, J. H., *Biochemistry*, **30**, 8124–8130 (1991).

60. Huang, S., Appleman, J. R., Tan, X., Thompson, P. D., Blakley, R. L., Sheridan, R. P., Venkataraghavan, R., and Freisheim, J. H., *Biochemistry*, **29**, 8063–8069 (1990).

61. Blakley, R. L., Piper, J. R., Maharaj, G., Appleman, J. R., Delcamp, T. J., Freisheim, J. H., Kulinsky, R. F., and Montgomery, J. A., *Eur. J. Biochem.*, **196**, 271–280 (1991).

62. Baccanari, D. P., Tansik, R. L., Joyner, S. S., Fling, M. E., Smith, P. L., and Freisheim, J. H., *J. Biol. Chem.*, **264**, 1100–1107 (1989).

63. Kaufman, B. T. and Gardiner, R. C., *J. Biol. Chem.*, **241**, 1319–1328 (1966).

64. Kaufman, B. T. and Kemerer, V. F., *Arch. Biochem. Biophys.*, **172**, 289–300 (1976).

64a. Mareya, S. M., and Blakley, R. L. unpublished results.

65. Webber, S. and Whiteley, J. M., *Arch. Biochem. Biophys.,* **236,** 681–690 (1985).

66. Fierke, C. A., Johnson, K. A., and Benkovic, S. J., *Biochemistry,* **26,** 4085–4092 (1987).

67. Beard, W. A., Appleman, J. R., Delcamp, T. J., Freisheim, J. H., and Blakley, R. L., *J. Biol. Chem.,* **264,** 9391–9399 (1989).

68. Filman, D. J., Bolin, J. T., Matthews, D. A., and Kraut, J., *J. Biol. Chem.,* **257,** 13663–13672 (1982).

69. Bystroff, C., Oatley, S. J., and Kraut, J., *Biochemistry,* **29,** 3263–3277 (1990).

70. Wu, Y.-D. and Houk, K. N., *J. Am. Chem. Soc.,* **109,** 2226–2227 (1987).

71. Donkersloot, M. C. A. and Buck, H. M., *J. Am. Chem. Soc.,* **103,** 6549–6554 (1981).

72. Bajoreth, J., Kitson, D. H., Fitzgerald, G., Andzelm, J., Kraut, J., and Hagler, A. T., *Proteins: Struct. Funct. Genet.,* **9,** 217–224 (1991).

73. Bajoreth, J., Kraut, J., Li, Z., Kitson, D. H., and Hagler, A. T., *Proc. Natl. Acad. Sci. USA,* **88,** 6423–6426 (1991).

74. Bajoreth, J., Kitson, D. H., Kraut, J., and Hagler, A. T., *Proteins: Struct. Funct. Genet.,* **11,** 1–12 (1991).

75. Huennekens, F. M. and Scrimgeour, K. G., Tetrahydrofolic acid—the coenzyme of one-carbon metabolism, in *Pteridine Chemistry,* Pfleiderer, W. and Taylor, E. C., Eds., MacMillan, Oxford, England, 1964.

76. Stuart, A., Wood, H. C. S., and Duncan, D., Pteridine derivatives. Part X. *J. Chem. Soc., C,* 285–288 (1966).

77. Vonderschmitt, D. J., Vitols, K. S., Huennekens, F. M., and Scrimgeour, K. G., *Arch. Biochem. Biophys.,* **122,** 488–493 (1967).

78. Poe, M., *J. Biol. Chem.,* **248,** 7025–7032 (1973).

79. Poe, M., *J. Biol. Chem.,* **252,** 3724–3728 (1977).

80. Cocco, L., Groff, J. P., Temple, C., Jr., Montgomery, J. A., London, R. E., Matwiyoff, N. A., and Blakley, R. L., *Biochemistry,* **20,** 3972–3978 (1981).

81. Maharaj, G., Selinsky, B. S., Appleman, J. R., Perlman, M., London, R. E., and Blakley, R. L., *Biochemistry,* **29,** 4554–4560 (1990).

82. Uchimaru, T., Tsuzuki, S., Tanabe, K., Benkovic, S. J., Furukawa, K., and Taira, K., *Biochem. Biophys. Res. Commun.,* **161,** 64–68 (1989).

83. Gready, J. E., *Biochemistry,* **24,** 4761–4766 (1985).

84. Selinsky, B. S., Perlman, M. E., London, R. E., Unkefer, C. J., Mitchell, J., and Blakley, R. L., *Biochemistry,* **29,** 1290–1296 (1990).

85. Chen, Y. Q., Kraut, J. Blakley, R. L., and Callender, R., *Biochemistry,* **33,** 7021–7026 (1994).

86. Stone, S. R. and Morrison, J. F., *Biochemistry,* **23,** 2753–2758 (1984).

87. Morrison, J. F. and Stone, S. R., *Biochemistry,* **27,** 5499–5506 (1988).

88. Howell, E. E., Villafranca, J. E., Warren, M. S., Oatley, S. J., and Kraut, J., *Science,* **231,** 1123–1128 (1986).

89. London, R. E., Howell, E. E., Warren, M. S., Kraut, J., and Blakley, R. L., *Biochemistry,* **25,** 7229–7235 (1986).

90. Murphy, D. J. and Benkovic, S. J., *Biochemistry,* **28,** 3025–3031 (1989).

91. Taira, K., Chen, J.-T., Fierke, C. A., and Benkovic, S. J., *Bull. Chem. Soc. Jpn.,* **60,** 3025–3030 (1987).

92. Blakley, R. L., Appleman, J. R., Freisheim, J. H., and Jablonski, M. J., *Arch. Biochem. Biophys.,* **306,** 501–509 (1993).

93. Cheung, H. T. A., Birdsall, B., Frenkiel, T. A., Chau, D. D., and Feeney, J., *Biochemistry,* **32,** 6846–6854 (1993).

94. Holbrook, J. J. and Gutfreund, H., *FEBS Lett.,* **31,** 157–169 (1973).

95. Dubrow, R. and Pizer, L. I., *J. Biol. Chem.,* **252,** 1539–1551 (1977).

96. Wagner, C. R., Thillet, J., and Benkovic, S. J., *Biochemistry,* **31,** 7834–7840 (1992).

97. Rosenberg, I. H. and Selhub, J., Intestinal absorption of folates, in *Folates and Pterins,* Vol. 3, *Nutritional, Pharmacological and Physiological Aspects,* Blakley, R. L. and Whitehead, V. M., Eds., J Wiley, New York, 1986, pp. 147–176.

98. Brown, G. M., Biosynthesis of H_4biopterin and related compounds, in *Chemistry and Biology of Pteridines 1989. Pteridines and Folic Acid Derivatives,* Curtius, H.-Ch., Ghisla, S., and Blau, N., Eds., de Gruyter, New York, 1990, pp. 199–212.

99. Thibault, T., Koen, M. J., and Gready, J. E., *Biochemistry,* **28,** 6042–6049 (1989).

TYROSINE HYDROXYLASE

By SEYMOUR KAUFMAN, *Laboratory of Neurochemistry, National Institute of Mental Health, Bethesda, Maryland*

CONTENTS

I. Introduction

The discovery of the essential role of unconjugated pterins in the enzymatic hydroxylation of phenylalanine (1, 2), and the identification of the naturally occurring hydroxylation cofactor in liver as tetrahydrobiopterin (BH_4) (3), facilitated the subsequent characterization of both tyrosine hydroxylase (TH) and tryptophan hydroxylase (TPH).

Advances in Enzymology and Related Areas of Molecular Biology, Volume 70, Edited by Alton Meister.
ISBN 0-471-04097-5   © 1995 John Wiley & Sons, Inc.

With respect to TH, for example, it had been postulated by Blaschko in 1939 (4), that the first step in the biosynthesis of norepinephrine and epinephrine involved the conversion of tyrosine to 3,4-dihydroxyphenylalanine (dopa), followed by decarboxylation of the dopa to dopamine, which was then hydroxylated in the β position of the side chain and methylated on the amino group to form norepinephrine and epinephrine, respectively (Fig. 1). Although the evidence in favor of the pathway proposed by Blaschko continued to mount (5–8), and it is now widely accepted as the only quantitatively important one, the enzyme responsible for the ring hydroxylation

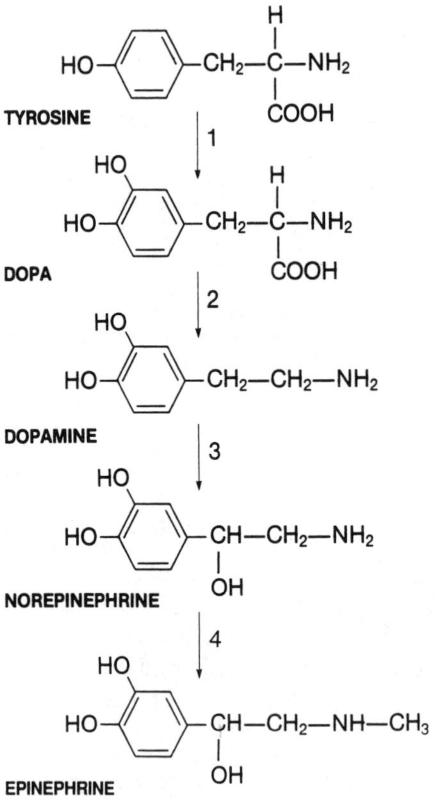

Figure 1. Biosynthetic pathway for the conversion of tyrosine to norepinephrine and epinephrine.

reaction remained elusive. Thus, long after the other three enzymes in the biosynthetic pathway had been characterized, including dopamine β-hydroxylase, the enzyme that is responsible for the side chain hydroxylation of dopamine (Fig. 1, Reaction 3) (9), "tyrosine hydroxylase" had still not been described.

In 1964, the enzymatic conversion of L-tyrosine to L-dopa was demonstrated in particles isolated from adrenal medulla (10–12), brain, and other sympathetically innervated tissues (12). The properties of this enzyme clearly distinguished it from tyrosinase, another enzyme that is capable of oxidizing tyrosine to dopa. With the latter enzyme, however, dopa is further oxidized to melanin [see (13) for a useful outline of the differences between tyrosinase and TH].

The sharpest distinction between TH and tyrosinase, a distinction that could be demonstrated even in crude fractions of the enzyme from adrenal medulla, is that TH shows an absolute and specific requirement for a tetrahydropterin (10–12). Dopa, for example, which is capable of stimulating the activity of tyrosinase, is unable to substitute for a tetrahydropterin (10, 11), and, in fact, is a potent inhibitor of TH (12). As far as the pterins are concerned, not only was the naturally occurring pterin cofactor BH_4 shown to be highly active, but significantly, it was demonstrated that it could function catalytically (10, 11). Furthermore, the marked stimulation of the conversion of tyrosine to dopa by NADPH and dihydropteridine reductase (DHPR) showed that the tetrahydropterin was being utilized during the hydroxylation reaction (10, 11). All of these general characteristics of the TH-catalyzed reaction indicated that the role of the tetrahydropterin in this hydroxylation system would prove to be identical to the one previously established in the phenylalanine hydroxylating system (14, 15), an expectation that was fulfilled by subsequent work.

Surprisingly, it was not until 1971 that the precise nature of the reaction catalyzed by TH was determined. Studies of the stoichiometry of the reaction in the presence of BH_4 showed that the reaction proceeded according to Eq. (1), where q-BH_2 stands for quinonoid dihydrobiopterin (16):

$$L\text{-tyrosine} + BH_4 + O_2 \rightarrow L\text{-dopa} + q\text{-}BH_2 + H_2O \qquad (1)$$

Following the demonstration that the 4a-hydroxytetrahydropterin is the primary pterin product formed during the PAH-catalyzed hydroxylation of phenylalanine (17, 18), this product was also shown to be found during the TH-catalyzed conversion of tyrosine to dopa. (For a further discussion of this aspect of the reaction, see Section XI and Fig. 7.)

II.　Intracellular Localization and Regional Distribution

As already mentioned, TH is present in brain, adrenal medulla, and sympathetically innervated tissues. Its intracellular location in these tissues, however, has been the subject of controversy. Udenfriend and co-workers (12) reported that in guinea pig brain all of the activity, and in beef adrenal medulla, most of the activity, was particle bound. They further concluded that the hydroxylase found in the soluble fraction of adrenal medulla was originally present in particles from which it had been released by the homogenization process, a conclusion that was strengthened by the finding that about 90% of the activity was sedimentable after centrifugation for 1 h at 144,000 g. These findings appeared to support the proposal that all of the enzymes involved in catecholamine biosynthesis are localized within a catecholamine-containing granule (19).

There is a growing body of evidence that contradicts the conclusion that TH is present intracellularly within a granule (20, 21). For the enzyme from both adrenal medulla and nerve tissue, it has been found that a major fraction of the activity is present in the high-speed supernatant fraction and that the distribution of hydroxylase activity is a function of the composition of the homogenization medium; homogenization in isotonic KCl leads to more enzyme in the supernatant fraction than does homogenization in isotonic sucrose (21). At least part of the uncertainty about the intracellular localization of TH can be traced to the tendency of the enzyme from bovine adrenal medulla, and probably from other tissues as well, to aggregate and to adsorb to subcellular organelles (21).

Although there is now a near consensus in favor of the view that the high molecular weight form of the enzyme in adrenal medulla is an artifact that is formed during cell disruption and that it is not in equilibrium with the native, nonaggregated form (22), that conclusion does not apply to brain where the enzyme appears to exist in two distinct forms, a soluble and a membrane-bound form. In fact, in the earliest studies on the brain enzyme, all of the activity was reported to be particle bound (12, 23). Later, it was shown that a soluble form is also present and that the distribution between soluble and particulate forms is brain-region specific; in brain areas that are rich in cell bodies of dopaminergic neurons, such as the *substantia nigra,* the soluble form predominates, whereas in those areas that are rich in nerve endings or axon terminals from these dopaminergic neurons, such as the caudate nucleus and putamen, the insoluble or particulate form predominates (24, 25). Based on this distribution, it was postulated that TH is synthesized in a soluble form in the perikarya of dopaminergic neurons and that it becomes more tightly bound as it moves along the axon to the nerve endings (24).

A similar partitioning of TH between soluble and insoluble forms occurs in noradrenergic neurons. Thus, in what is designated the dorsal noradrenergic pathway in which the cell bodies are enriched in the *locus coeruleus,* most of the enzyme is present in the soluble fraction, whereas in the hippocampus-cortex, an area rich in noradrenergic nerve endings, most of the enzyme is particle bound (26).

These conclusions were confirmed and extended with immunochemical techniques. With the use of a specific antibody to TH, the enzyme was localized by electron microscopy within the cytoplasm of noradrenergic neurons of the nucleus *locus coeruleus,* whereas in the axons and dendrites, the enzyme appeared as fiberlike strands aligned parallel with the plasma membrane, attached to structures that were tentatively identified as neurotubules (27).

As will be discussed in greater detail later, some of the kinetic properties of the soluble and particulate forms of brain TH are different, the most dramatic difference being that the K_m of the particulate form for 6,7-dimethyltetrahydropterin ($DMPH_4$) is only one-fifth and the V_{max} is twice as great as that of the soluble form (25). It has also been postulated that the TH activity may be regulated by alterations in the physical state of the enzyme (25).

There have been several studies of the regional distribution of TH in rat brain (28–30). There is general agreement that regions rich in dopaminergic cell bodies (*substantia nigra*) and nerve terminals (striatum) have high TH specific activities. As for regions rich in cell bodies of noradrenergic neurons, such as *locus coeruleus* and *pons medulla,* they have much lower activities than that of *substantia nigra,* but not as low as regions rich in noradrenergic nerve terminals, such as cortex, cerebellum, and hippocampus. The hypothalamus, the region with the highest density of noradrenergic nerve terminals, has relatively high TH activity (Table I).

Significant correlations between the distribution of TH and that of "cofactor-active tetrahydropterin" have been found (29). The same trend was noted by Levine et al. (30), when pterin cofactor activity was plotted against the sum of both TPH and TH activities (with the exception of hypothalamus, which had a much higher ratio of pterin cofactor to the sum of the two hydroxylases). In view of the subsequent demonstration that another BH_4-dependent enzyme, nitric oxide synthase, is enriched in cerebellum (31), it is of interest that this brain area in rats has one of the lowest levels of the pterin cofactor (30), a finding that can be reconciled by the very low K_m of that enzyme for BH_4 (32). In contrast with the correlation between the levels of the aromatic amino acid hydroxylases and pterin cofactor, no correlation across different brain regions was found between TH activity and the activity of DHPR (29). In accord with this result, Craine et al. (33) previously found that the distribution of the latter enzyme in various regions of rat brain was relatively homogeneous.

III. Methods of Assay

The earliest assays of TH were based on the ability of alumina to adsorb catechols such as dopa, but not tyrosine. After separation of the radioactive tyrosine from the dopa, the dopa was eluted from

TABLE I
Regional Distribution of Tyrosine Hydroxylase in Rat
Brain[a]

Brain Region	Tyrosine Hydroxylase (pmol/mg min ± SEM)
Cortex	2.2 ± 0.5
Hippocampus	2.5 ± 0.3
Cerebellum	2.8 ± 0.7
Pons medulla	21.2 ± 1
Striatum	598 ± 70
Hypothalamus	69.8 ± 5
Tegmentum	173 ± 17

[a] Data from Levine et al. (30).

the alumina with acid and the amount of radioactivity was determined (10, 12). To assay crude preparations of the enzyme, inhibitors of monoamine oxidase and dopa decarboxylase were added to the reaction mixtures (10, 12).

In their initial report, Nagatsu et al. (12) used an excess of tetrahydrofolate and Fe^{2+} in their standard assay. In contrast, Brenneman and Kaufman (10) used catalytic amounts of $DMPH_4$ together with those components that had been previously shown to play a role in the phenylalanine hydroxylating system, that is, reduced pyridine nucleotide and DHPR to regenerate the $DMPH_4$ was also included.

A tritium-release assay of TH was developed that takes advantage of the fact that during the enzymatic conversion of L-(3,5-^{3}H)tyrosine to L-dopa, approximately one-half of the tritium is released into the medium. Dowex-50 columns were used to adsorb the radioactive tyrosine and dopa, and a sample of the effluent containing the tritiated water was counted (12). The same principle had been used previously to develop a tritium-release assay for tyrosinase (34).

Although this method of separating the amino acids from the ^3H_2O worked well with partially purified enzyme preparations, in crude systems significant amounts of the tyrosine can be metabolized to products that are not retained by Dowex-50. To remedy this source of error, Karobath (35) suggested that the Dowex-50 eluates be passed through a Dowex-1 column to remove radioactive acidic products.

A modification of the tritium-release assay that does not suffer from the just-mentioned disadvantage of the ion-exchange separation of the amino acids and ^3H_2O has been described (16). In this version, the ^3H_2O is separated from the amino acids by lyophilization of the water. Since this method would not be affected by further metabolism of the hydroxylated product, dopa (unless it was metabolized to water), it can be applied to crude systems and even to whole cells. Another useful variation of the tritium-release assay uses charcoal to adsorb the isotopic substrate and its metabolites (36).

An assay for TH that has been widely used for determining the enzyme's activity in intact cells and tissue extracts is based on the hydroxylase-catalyzed conversion of L-1-^{14}C tyrosine to L-1-^{14}C-dopa. The dopa that is formed is then decarboxylated by aromatic amino acid decarboxylase and the $^{14}CO_2$ collected and its radioactivity determined (37). This coupled hydroxylase-decarboxylase method, which was first introduced for the assay of TPH (38), works because the K_m values of the decarboxylase for the hydroxylated products (i.e., for 5-hydroxytryptophan and dopa for TPH and TH, respectively, are much lower than the K_m values for the amino acid substrates. As is true for any coupled enzymatic assay, when the effects on TH of drugs or inhibitors are being tested, it must be established that the substances being tested are not inhibiting the decarboxylase rather than the hydroxylase.

With the application of high-performance liquid chromatography (HPLC) with electrochemical detection in the determination of dopamine and norepinephrine in brain (39–42), it was only a matter of time before these methods would be applied to the assay of TH. Several of these HPLC assays have been described (43, 44).

With partially purified preparations of TH, many of the assays that had been previously developed for the assay of phenylalanine hydroxylase (PAH) (45), have been used, including the coupled spectrophorometric assay in which the hydroxylation reaction is coupled to the DHPR-catalyzed NADH or NADPH-mediated reduction of the quinonoid dihydropterin back to the tetrahydro level (16), tyrosine-dependent oxidation of BH_4 (or another active tetrahydropterin) monitored at 330–340 mμ (16), and tyrosine-dependent oxygen uptake (16).

IV.   Substrate Specificity

A.   TETRAHYDROPTERIN COENZYME

TH's requirement for a tetrahydropterin appears to be absolutely specific. The following compounds have been tested as possible substitutes for the reduced pterin and were found to be completely inactive at a final concentration of 0.1 mM: ascorbate (10, 12), cysteine (10), glutathione, $FeCl_2$ (10, 12), flavin mononucleotide (10), FAD (10), and dopa (10, 12).

Within the pterin series, unconjugated tetrahydropterins are far more active than tetrahydrofolate (10, 12). Although the latter compound was reported to have good cofactor activity, especially in the presence of Fe^{2+} (12), subsequent work showed that most, if not all, of this activity is the result of contamination of the tetrahydrofolate with 6-methyltetrahydropterin (46).

As far as specificity for unconjugated pterins is concerned, more detailed studies have been carried out with the enzyme from adrenal medullary tissue than with the brain enzyme. With partially purified preparations of the native enzyme from adrenals, it was shown that TH resembles PAH (2), in that it is more active with 6-methyltetrahydropterin ($6MPH_4$) than with $DMPH_4$ (10). An unsubstituted N5 position is necessary for cofactor activity, as is a 2-amino or a 4-hydroxy substituent; a pteridine with neither of these substituents is completely inactive (47).

A comparison of the cofactor activities of BH_4, $6MPH_4$, and $DMPH_4$ with both the particulate form of the bovine adrenal enzyme and with the form that has been solubilized by limited proteolysis with chymotrypsin (referred to as the "solubilized" enzyme), together with data for the bovine caudate enzyme is shown in Table II. As can be seen for both adrenal and the brain TH, the apparent K_m for BH_4 is lower than for $6MPH_4$ or $DMPH_4$ (16, 48). It should be noted, however, that the differences in K_m values for BH_4 and these synthetic pterins with TH are not nearly as marked as they are for rat liver PAH where the K_m values for $6MPH_4$ and $DMPH_4$ are 20 and 30 times greater, respectively, than the K_m for BH_4 (45). The relative velocities with BH_4, $6MPH_4$ and $DMPH_4$ are 1.0, 0.9, and 0.53 for the solubilized enzyme and 1.0, 0.65, and 0.45 for the particulate adrenal enzyme (16, 48).

TABLE II
Michaelis Constants (K_m) of Bovine Adrenal and Caudate Tyrosine Hydroxylase for Tyrosine and Pterin Cofactors[a]

Cofactor	Solubilized Adrenal Enzyme (mM)	Particulate Adrenal Enzyme (mM)	Bovine Caudate Enzyme (mM)
Tyrosine (BH_4)	0.015	0.004	0.010
Tyrosine ($6MPH_4$)	0.07	0.04	0.075
Tyrosine ($DMPH_4$)	0.20	0.10	0.075
BH_4	0.10	0.10	0.22
$6MPH_4$	0.30	0.30	0.80
$DMPH_4$	0.30	0.30	1.00

[a] Data from Kaufman (64).

Detailed studies of the effect on coenzyme activity of structural changes in the 6-dihydroxypropyl side chain of BH_4 have been carried out. As can be seen from the data summarized in Table III, although there is considerable variation in the actual values across different studies, they all show, in agreement with the data in Table II, that the K_m for BH_4 is lower than the K_m of any of the synthetic analogs, and, the K_m for the natural isomer of BH_4, $(6R)$-BH_4, is lower than the K_m of the unnatural $6S$ isomer. It should be emphasized that because TH preparations of different specific activities were used in these studies, comparisons of the V_{max} values are only valid within a given study and not across the different studies.

In contrast to the relative velocities (not V_{max} values) mentioned above, however, the V_{max} value for $6MPH_4$ was found by Numata (Sudo) et al. (49), to be greater than that for L-*erythro* BH_4. It is of interest that although all of the isomers of BH_4 have lower K_m values than $6MPH_4$, an indication that a dihydroxypropyl side chain is im-

TABLE III

Pterin K_m and V_{max} Values for Bovine Adrenal Tyrosine Hydroxylase

Pterin	K_m (μM)	Maximum Velocity (nmol/mg/min)
BH_4	19	11.7[a]
D-Erythro-6-(1′,2′-dihydroxypropyl)-tetrahydropterin	37	6.14[a]
L-Threo-6-(1′,2′-dihydroxypropyl)-tetrahydropterin	58	8.15[a]
D-Threo-6-(1′,2′-dihydroxypropyl)-tetrahydropterin	21	10.8[a]
$6MPH_4$	103	13.9[a]
$DMPH_4$	87	9.6[a]
6-L-Hydroxyethyl-tetrahydropterin	57	5.9[b]
6-D-Hydroxyethyl-tetrahydropterin	89	5.8[b]
6-L-Dihydroxyethyl-tetrahydropterin	218	7.3[b]
6-D-Dihydroxyethyl-tetrahydropterin	234	4.1[b]
$6MPH_4$	115	10.7[b]
6-Hydroxymethyl-tetrahydropterin	230	9.8[b]
$(6R)$-BH_4	74	143[c]
$(6S)$-BH_4	522	120[c]

[a] Data from Numata (Sudo) et al. (49).

[b] Data from Kato et al. (50).

[c] Data from Oka et al. (65).

portant for good binding, neither a hydroxyethyl, a dihydroxyethyl, nor a hydroxymethyl substituent at position 6 of the tetrahydropterin significantly decreased the K_m value compared to that found for 6MPH$_4$ (50). Indeed, the latter two hydroxylated side chains increased the K_m value compared to 6MPH$_4$.

In studies carried out with the enzyme from bovine striatum, it was found that the K_m values for both BH$_4$ and 6MPH$_4$ vary markedly with pH (51). As can be seen in Table IV, the K_m values for both pterins are about 10 times larger at pH 6.0, near the optimum pH for the enzyme, than they are at pH 7.3. Very similar changes in both K_m and V_{max} values for 6MPH$_4$ were also reported for the enzyme from rat striatum (51).

Although the changes with pH were not as great, Lazar et al. (52) found changes in the same direction for BH$_4$, that is, both the K_m and V_{max} values were lower at pH 7 than at 6. At the latter pH, their values for the K_m of BH$_4$ (345 μM), and for 6MPH$_4$ (480 μM), were in excellent agreement with those shown in Table IV. They found, however, that at pH 6.0, the V_{max} for 6MPH$_4$ was 1.7 times larger than the V_{max} for BH$_4$. In sharp contrast to the general agreement between the two studies just discussed, Kuczenski (53) reported that with rat striatal TH the K_m for DMPH$_4$ was much lower at pH values below 5.7 than it was above pH 6.

These pH studies reveal another noteworthy difference in properties between TH from adrenals and from brain. In contrast to the marked change with pH of the K_m values for the tetrahydropterins shown by the bovine striatal enzyme (Table IV), it has been reported that for pure TH from rat adrenals, the K_m for 6MPH$_4$ does not vary with pH over the range of pH 5.2–8.0 (54). An indication that the

TABLE IV
Pterin K_m Values for Bovine Striatal Tyrosine Hydroxylase[a]

Tetrahydropterin	K_m (μM)		V_{max} (nmol/min/mg)	
	pH 6.0	pH 7.3	pH 6.0	pH 7.3
BH$_4$	590	30	3.6	0.1
6MPH$_4$	720	64	3.3	0.3

[a] Data from Pollock et al. (51).

pattern of change in K_m values of TH for tetrahydropterins as a function of pH is still more variable comes from the report that, for the enzyme from rat pheochromocytoma, the K_m of 6MPH$_4$ varies in the opposite way from the way it varies with TH from bovine striatum. Specifically, the K_m *increases* markedly on going from pH 6.0 to 7.2: The K_m at pH 6 was reported to be 0.20 mM, whereas at pH 7.0 it was found to be 0.92 mM (55).

Bailey et al. (56) studied the effect of variations in the hydrophobicity of the side chain at the 6 position of the pterin cofactor on the affinity (i.e., K_m values) of these compounds for a crude preparation of bovine striatal TH. With enzyme that was not specifically phosphorylated (designated unphosphorylated) and was studied at pH 7.15, they found that with a series of pterins where the dihydroxypropyl substituent at position 6 of BH$_4$ was replaced by substituents of increasing hydrophobicity, specifically, 6-hydroxymethyl, 6-methyl, 6-methoxy, 6-ethyl, 6-phenyl, and 6-cyclohexyl, the K_m values generally decreased with increasing hydrophobicity. These values ranged from a high of about 3–5 mM for 6MPH$_4$ and (6R)-BH$_4$ to a low of 0.3 mM for 6-cyclohexyltetrahydropterin. Relative V_{max} values also decreased with increasing hydrophobicity of the six-substituent, but not as markedly as did the K_m values. Therefore, the relative V_{max}/K_m values of some of these pterins with more hydrophobic side chains, such as the 6-cyclohexyl and 6-phenyl compounds, were 6–10 times greater than the value for (6R)-BH$_4$.

As will be discussed in greater detail in Section VI(C), TH can be activated by phosphorylation by a variety of kinases, including cAMP-dependent protein kinase (PKA). One of the expressions of the activated state is a marked decrease in the K_m for pterin cofactors, as was first demonstrated with 6(R,S)-BH$_4$, 6MPH$_4$, DMPH$_4$ (57), and (6R)-BH$_4$ (58). In their study of pterins with side chains of increasing hydrophobicity, Bailey et al. (56) also studied the interaction of these compounds with phosphorylated TH (phosphorylated by the action of PKA). In confirmation of the earlier findings (57, 58), they found that phosphorylation of the hydroxylase led to a large decrease in the K_m for all of the pterins in their series, the effect ranging from about a 2000-fold decrease for (6R)-BH$_4$ and 6-ethyltetrahydropterin to a 50-fold decrease for 6MPH$_4$. With the phosphorylated enzyme, there was also, with most of the pterins, an inverse correlation between K_m and increasing hydrophobicity of the 6 substituent. There were, however, exceptions to this relationship with both (6R)-BH$_4$ (K_m = 3 μM) and 6-hydroxymethyltetrahydropterin (K_m = 20 μM) having lower values than the more hydrophobic 6MPH$_4$ (K_m = 60 μM).

Based on their results, Bailey et al. (56) postulated that binding of tetrahydropterins to TH involves both hydrophilic regions of the protein (interacting with the hydroxyl groups of the pterin side chain) and hydrophobic regions of the protein (recognizing the hydrophobic groups of the pterin side chain).

Parenthetically, it should be noted that the K_m values of 5 mM for (6R)-BH$_4$ and 3 mM for 6MPH$_4$ reported by Bailey et al. (56), are almost two orders of magnitude larger than those reported for these compounds with partially purified bovine striatal TH at pH 7.3 (see Table IV). The reason for these discrepancies are not known but may be related to the use of what the authors characterized as a "minimally purified TH preparation" in the study that gave the very high values (56), compared to a partially purified preparation in the other study (51). In this regard, a K_m value for BH$_4$ of 200 μM at pH 7.0 was reported for highly purified bovine striatal TH (52), a value that, according to published data (51), would probably be about 100 μM at pH 7.3 and, therefore, not too different from the value of 30 μM shown in Table IV. Coherent with the possibility that this unusually high value for the K_m for BH$_4$ may be due to the use of rather crude TH preparations, a K_m for (6R)-BH$_4$ nearly as high (2.3 mM at pH 7.0) was obtained when unfractionated rat striatal extracts were used as the source of the enzyme (58). This latter value, however, was based on hydroxylase assays that were characterized by a lag period, lasting longer than 10 min, that became progressively more pronounced at BH$_4$ concentrations greater than 100 μM. It may be difficult to extract a meaningful K_m value from such peculiar kinetics.

An important aspect of the specificity of tetrahydropterins that is not revealed by the mere compilation of K_m and V_{max} values is the question of whether the stoichiometry of the hydroxylation reaction is the same in the presence of all of the pterins that have been tested. In most studies of the structure-activity relationship of tetrahydropterins, this question has been largely neglected. Earlier, it was shown that the phenomenon of partial uncoupling of hydroxylation from electron transfer, first discovered with PAH (59, 60), also occurs with adrenal TH (16). The structural requirements necessary to elicit uncoupling, however, differ with the two hydroxylases.

The most striking difference is that with $DMPH_4$, tight coupling is observed with PAH (60), whereas there is loose coupling with TH (16). Table V summarizes the results obtained with various tetrahydropterins and highly purified solubilized adrenal TH. As mentioned previously, there is tight coupling with BH_4 as the cofactor (dopa/dihydrobiopterin = 1.0).

From this limited study of pterin analogs, the presence of a methyl substituent in position 7 of the pterin ring appears to lead to anomalous stoichiometry with TH. By contrast, with PAH, it is the absence of a substituent in position 6 that appears to be the structural requirement for the anomalous stoichiometry (60). This conclusion is based not only on the different results obtained for the two enzymes with $DMPH_4$, but also on those obtained with the unsubstituted pterin, PH_4; with PAH, the hydroxylation reaction with this pterin gives an anomalous ratio (60), whereas the ratio is close to 1.0 with the TH-catalyzed reaction. The significance of these different structural requirements exhibited by the two hydroxylases is not clear. The possible mechanistic interpretation of the phenomenon of uncoupling will be discussed later.

B. AMINO ACID SUBSTRATE

The adrenal enzyme was originally thought to be absolutely specific for L-tyrosine (12); none of the following compounds was active: D-tyrosine, tyramine, DL-m-tyrosine, or L-trytophan (12). The first indication that the adrenal enzyme is not absolutely specific for

TABLE V
Stoichiometry of Tyrosine Hydroxylation with Model Cofactors[a]

Tetrahydropterin	Dihydropterin/ Dopa	Range	Number of Determinations
$6MPH_4$	1.10 ± 0.05	(1.05–1.12)	6
$7MPH_4$	1.60 ± 0.05	(1.56–1.62)	4
PH_4	1.15 ± 0.05	(1.12–1.15)	4
$DMPH_4$	1.75 ± 0.15	(1.61–2.0)	10

[a] Data from Shiman et al. (16).

L-tyrosine was the finding that L-phenylalanine could serve as a substrate (61). With a relatively crude enzyme preparation and $DMPH_4$ as the cofactor, the rate of hydroxylation of L-phenylalanine was about 5% of the rate of hydroxylation of L-tyrosine under comparable conditions. It was also shown that the initial product of phenylalanine hydroxylation was L-tyrosine.

The low rate of phenylalanine hydroxylation by TH in the presence of $DMPH_4$ indicated that this reaction was probably of minor physiological significance. A more detailed study of the phenylalanine hydroxylating ability of TH, however, led to the surprising finding that in the presence of BH_4 the rate of phenylalanine hydroxylation by both the highly purified, solubilized enzyme and the particulate enzyme from bovine adrenal medulla is equal to, or greater than, the rate of tyrosine hydroxylation (16). Furthermore, whereas the enzyme is sensitive to inhibition by excess tyrosine, excess phenylalanine does not inhibit. These results raised the possibility that *in vivo* L-phenylalanine could serve as a significant precursor of norepinephrine.

The original conclusion about the sharp amino acid specificity of TH was modified still further by the studies of Tong et al. (62, 63). Using a crude preparation of the adrenal enzyme, they showed that it could catalyze the conversion of L-*m*-tyrosine to dopa at about 50% the rate at which L-tyrosine is converted to dopa. In addition, they reported that the preparation could catalyze the conversion of L-phenylalanine to *m*-tyrosine at about 15% of the rate at which L-phenylalanine is converted to tyrosine (62). Although it is not certain that TH is the responsible enzyme, the sensitivity of the *m*-hydroxylation reaction to the TH inhibitors, 3-iodo-tyrosine and α-methyltyrosine, supports the conclusion that the same enzyme is catalyzing the hydroxylation of phenylalanine in the para and the meta positions. The conversion of phenylalanine to dopa via *m*-tyrosine is a potential alternate pathway for dopa formation, although its *in vivo* significance relative to the phenylalanine–tyrosine–dopa pathway remains to be established.

A property of the pterin-dependent hydroxylases that is not fully appreciated is that the apparent K_m values for their substrates (both the amino acid substrate and oxygen) vary with the pterin used (45). The summary of K_m values shown in Table II illustrates this point for tyrosine with bovine adrenal and caudate TH (48, 64). For both the purified, solubilized, and the particulate adrenal enzyme, as well as for the brain enzyme, the K_m value for tyrosine is much lower in the presence of BH_4 than it is in the presence of $DMPH_4$. Excess tyrosine inhibits TH in the presence of either BH_4 or $6MPH_4$, the inhibition being more pronounced with BH_4 (50% inhibition at ~100 μM of tyrosine). For this reason, the listed K_m values for tyrosine in the presence of this pterin are only approximate. Tyrosine does not inhibit in the presence of $DMPH_4$ (16).

The finding that the K_m of tyrosine varies with the structure of the pterin cofactor has been confirmed with TH purified from bovine adrenal medulla (49, 50, 65), and with the enzyme from brain (52, 57, 64, 66–68).

The possibility that phenylalanine might serve as an important precursor of norepinephrine *in vivo* was supported by results of studies in which radioactive phenylalanine was injected into the brains of rats (68, 69), as well as those in which the conversion of phenylalanine to catecholamine was demonstrated in isolated nerve endings (synaptosomes) from brains of various species (70). In all of the animals studied, the rate of catecholamine formation from phenylalanine was 40–50% that from tyrosine. In another study in which phenylalanine and tyrosine were compared (71), it was shown that the behavior of phenylalanine as a substrate for isolated TH was significantly different from that in synaptosomes. With the isolated enzyme, although the rate of conversion of phenylalanine to tyrosine was comparable to the rate of conversion of tyrosine to dopa, the rate of conversion of phenylalanine to dopa was only about 10% that from tyrosine. As pointed out (71), the relatively low rate of dopa formation from phenylalanine is probably due to competition between the phenylalanine and the newly formed tyrosine for the active site on the hydroxylase. By contrast, with synaptosomes the rate of dopa formation from phenylalanine was 25% (71)–43% (70) the rate from tyrosine and it was shown that the tyrosine that is formed from the phenylalanine within the synaptosome does not equilibrate with the pool of added tyrosine (67).

The original finding that in the presence of BH_4 phenylalanine is an excellent substrate for isolated TH (16) has been widely replicated (72, 73), and, as mentioned above, it has also been demonstrated *in vivo* and in isolated synaptosomes. In addition, it has been shown to occur in bovine adrenal chromaffin cells (74). In view of this consensus, the report that pure TH isolated from cultured rat pheochromocytoma (PC12) cells was unable to utilize phenylalanine as a substrate (75), was startling. This negative result, however, could not be reproduced by Ribeiro et al. in 1991 (76) who found that TH purified to homogeniety from cultured PC12 cells does catalyze the hydroxylation of phenylalanine. The same was true for the enzyme isolated from PC18 cells, a subclone of PC12. Indeed, as originally reported for bovine adrenal TH (16), and rat brain TH (71), at substrate concentrations greater than 100 μM, the BH_4-dependent rate of phenylalanine hydroxylation was found to exceed the rate of tyrosine hydroxylation (76). The reason for the failure of Dix et al. (75) to detect phenylalanine hydroxylation by PC12 is unknown.

As a postscript to this discussion of the phenylalanine hydroxylating activity of TH, it should be added that, in contrast to the original negative result reported for TH from cultured PC12 cells (75), Andersson et al (77) recently showed that phenylalanine is a good substrate for TH isolated from tumors of rat pheochromocytoma cells. To explain the earlier failure of Dix et al. (75) to detect this activity in TH from cultured PC12 cells, Andersson et al. (77) postulated that the enzyme from the implanted tumor cells was different in this respect from the enzyme from the cultured cells. Clearly, the demonstration that TH from cultured cells can utilize phenylalanine as a substrate (76), obviates the need for such a postulated difference.

The ability of TH to catalyze the hydroxylation of phenylalanine may assume great importance in hyperphenylalaninemia, where the amount of tyrosine entering the brain from the periphery may be limited by competition with high blood levels of phenylalanine (78).

In addition to phenylalanine, it has been reported that a variety of ring-substituted phenylalanine derivatives, such as 4-amino-phenylalanine, 4-methoxy-phenylalanine, 4-fluoro-phenylalanine, and 3-fluoro-phenylalanine, as well as tryptophan, can serve as substrates for TH (73). The V_{max} values for all of the phenylalanine derivatives were fairly close to those for phenylalanine and tyrosine. By contrast, with the exception of 4-fluorophenylalanine, the K_m values for all of these phenylalanine derivatives were larger than that for tyrosine. The K_m value for tryptophan was 20 times larger than the value for tyrosine and its V_{max}/K_m value was only about 2% that for tyrosine.

V. Properties of Tyrosine Hydroxylase

A. SIZE

Many of the general characteristics of adrenal TH were initially determined on preparations that were purified after the enzyme had been solubilized by limited digestion with chymotrypsin (16) or trypsin (79). Despite the fact that protease treatment had altered the size of the enzyme, the kinetic constants of the chymotrypsin-treated enzyme, as can be seen in Table II, appeared to be similar to those of the native enzyme. The molecular weight of the trypsin-solubilized enzyme was reported to be 34,000 (80), and that of the essentially pure chymotrypsin-solubilized enzyme was found to be similar, 33,500–34,700 (81).

B. AMINO ACID COMPOSITION

Native rat adrenal TH is a tetramer, M_r = 260,000, composed of four identical subunits, M_r = 59,000 (54). These results indicate that the product of limited proteolysis (M_r = 34,000) no longer forms dimers or tetramers. In this regard, the behavior of TH resembles that of rat liver PAH. With the latter enzyme, it has been shown that in its native form it can form tetramers (M_r = 200,000). By contrast, the highly active product of limited chymotrypsin treatment (M_r = 35,000) has lost the ability to form tetramers, although it can still form dimers (82).

As is true of most of the other properties of TH, the situation with respect to the size of the brain enzyme appears to be more complicated than that of the adrenal enzyme. In this context, it has been reported (83) that the *locus coeruleus* of rat brain, a region enriched in cell bodies of noradrenergic neurons, contains a high molecular weight (M_r = 200,000) form of the enzyme, whereas in the *substantia nigra* and the caudate nucleus, the areas enriched in cell bodies and nerve endings, respectively, of dopaminergic neurons, a low molecular weight form of the enzyme (M_r = 65,000) is present and is the major form in the latter brain region. Peripheral noradrenergic neurons in the superior cervical ganglion contain a form of the enzyme with an intermediate molecular weight (M_r = 130,000). Further complicating the picture is the finding that incubation of homogenates of *locus coeruleus* with RNase decreased the molecular weight of the hydroxylase from about 200,000 to 150,000

(83). These results indicate that at least part of the apparently high molecular weight of the enzyme in this region is due to its association with RNA, a finding that makes it difficult to decide whether TH molecules of different structure are present in these different regions of the brain.

Although TH purified from bovine striatum has been reported to be composed of two different types of subunits, $M_r = 60,000$ and 62,000 (52), only a single type of subunit was detected for the adrenal enzyme ($M_r = 59,000$) (54), and for the enzyme from rat pheochromocytoma ($M_r = 62,000$) (54).

The deduced amino acid composition of TH from rat PC12 cells (84), is shown in Table VI. The composition is similar to that determined by chemical analysis of the rat adrenal enzyme (54), with the exception that the measured amounts of aspartic acid and glutamic acid were considerably greater than the deduced values.

TABLE VI
Amino Acid Composition of Tyrosine Hydroxylase from PC12 Cells[a]

Amino Acid Residue	Residues per Mole Subunit
Ala	51
Cys	7
Asp	29
Glu	42
Phe	27
Gly	23
His	18
Ile	15
Lys	21
Leu	50
Met	3
Asn	6
Pro	32
Gln	23
Arg	36
Ser	42
Thr	20
Val	33
Trp	3
Tyr	17
Total residues	498

[a] Data from Grima et al. (84).

VI.   Regulation of Tyrosine Hydroxylase Activity

A.   AVAILIBILITY OF SUBSTRATES AND COENZYMES

It is rarely possible to unequivocally answer the question of whether the *in vivo* activity of an enzyme is limited by the availability of its substrates and coenzymes. From a previous analysis of this question based on a comparison of the K_m values of TH for tyrosine, oxygen, and BH_4 with the tissue concentrations of these compounds, it was concluded that in the unactivated state, it is likely that the activity of the enzyme is limited by low tissue levels of BH_4 [see (85) for details]. The relatively high K_m for the pterin coenzyme sets the stage for the enzyme to be activated by processes that decrease this K_m value. As will be discussed, many modes of activation are mediated, in part, by this change.

With respect to the other two substrates, tyrosine and oxygen, a similar type of analysis indicated that the brain enzyme usually operates at about 80% of saturation with both of these substrates (85).

B.   EFFECTS OF HEPARIN AND OTHER POLYANIONS

The properties of TH from rat brain have been reported to be dramatically and rather specifically altered by microgram quantities of the sulfated mucopolysaccharide, heparin (25). As discussed in Section II, the enzyme is thought to occur in brain in both soluble and particulate forms. Although the two forms have identical K_m values for tyrosine, the particulate enzyme has a higher V_{max} and a lower K_m for $DMPH_4$ than does the soluble enzyme. Another significant difference between the two forms of the enzyme is that the K_i for norepinephrine is much lower (K_i = 0.1 mM) for the particulate enzyme than for the soluble enzyme (K_i = 0.7 mM).

The addition of heparin to the soluble enzyme was found to alter all of these kinetic properties so that it resembles the particulate enzyme. The observation that heparin markedly decreases the K_m of the soluble enzyme for $DMPH_4$ is noteworthy because it is the first example of the potential regulation of the activity of this enzyme by a mechanism that involves a decrease in the K_m for the pterin cofactor. The latter value, as already discussed, appears to be high relative to tissue concentrations of BH_4. As subsequent studies showed, these changes in the kinetic properties of the enzyme on activation proved to be characteristic.

Based on their findings, Kuczenski and Mandell (25) suggested that the site of binding of TH on a membrane surface may contain a sulfated mucopolysaccharide and that intracellular binding of the enzyme at this site may be an important determinant of the regulatory properties of the enzyme.

In addition to this effect of heparin, these workers also observed that the soluble brain TH is stimulated by high concentrations (0.1 M ionic strength) of salt with sulfate ions being the most active. In contrast to the effect of heparin, sulfate ions did not alter the apparent K_m for DMPH$_4$, but only increased V_{max}. Following the work of Kuczenski, Musacchio, et al. (86) found that soluble adrenal TH can also be activated by heparin.

That the activating effect of heparin on TH might not be as specific as originally believed (25), but instead, might be an example of a more general activation by polyanions, was indicated by the finding that certain phospholipids, such as phosphatidyl-L-serine, can stimulate partially purified bovine caudate TH (66). Although this finding was prompted by the earlier demonstration that phospholipids can markedly stimulate rat liver PAH (82), the mechanism of the activation by phospholipids proved to be more analogous to that observed with heparin on crude brain TH than to that with lysolecithin on PAH.

Of the phospholipids tested with purified bovine caudate TH, phosphatidyl-L-serine was found to be the most active, maximum activation being achieved at 0.4 mM (66). The mechanism of activation by the phospholipid was shown to involve a three- to fourfold decrease in the K_m of the pterin coenzyme, the decrease being demonstrated with DMPH$_4$, 6MPH$_4$, and BH$_4$ (66). This result with BH$_4$ was significant because it was the first demonstration that TH can be activated by a decrease in the K_m of the naturally occurring pterin cofactor, BH$_4$; neither the K_m for tyrosine nor the maximum velocity is affected by the phospholipid.

Although heparin has been reported to lead to a slight shift in the pH optimum of crude, soluble rat brain TH from pH 5.9 to 6.2 (25), this effect is not observed with the bovine caudate enzyme. Phosphatidyl-L-serine, however, caused a marked shift in the optimum pH from pH 6.0–6.1 to 6.6–6.8, so that the stimulation by the phospholipid at pH 6.8 is almost twice that at pH 6.1 (66).

A similar stimulation of TH by phospholipids, which also involves a decrease in the K_m of the pterin cofactor, has been demonstrated with a partially purified preparation of the enzyme from rabbit adrenals (87). In this study, anionic phospholipids were found to be active, whereas compounds without a net negative charge, for example, phosphatidyl choline and phosphatidyl ethanolamine, were not active; phosphatidylinositol was more active than phosphatidyl-L-serine. Just as with the stimulation of TH by heparin, it is not known whether the stimulation by phospholipids is significant for the physiological regulation of the activity of this enzyme.

The stimulation of TH by phospholipids was confirmed with a partially purified enzyme preparation from rat striatal synaptosomes (88). The results obtained were essentially the same as those described earlier for the bovine caudate enzyme (66). Slight stimulation of the rat brain enzyme by the nonanionic phospholipid, phosphatidyl ethanolamine, was observed, which contrasts with the lack of stimulation of the rabbit adrenal enzyme by this phospholipid (87).

The idea that the activation of TH by either heparin or anionic phospholipids might merely be specific examples of a more general activation by polyanions was supported by the finding that many polyanions, including polyacrylic acid, polyvinyl sulfuric acid (67), and polyglutamic acid (66, 67), can activate partially purified bovine caudate TH. Many different salts were found to both activate the enzyme [confirming the original finding (25)] and to prevent the activation by polyanions (67). Activation by univalent salts depended on the nature of both the cation and the anion used and was reversible (67). That the activation of the enzyme by a polyanion such as heparin is a result of a direct interaction between the two molecules was shown by the observation that TH is both bound to and activated by heparin-substituted Sepharose (67). This demonstration of direct binding led to the application of binding and elution of the enzyme from heparin-substituted Sepharose as a useful step in its purification (51).

These studies led to an interpretation of the heparin activation of TH that was radically different from the one expressed by Kuczenski and Mandell (25). These investigators assumed that the heparin effect was an example of "allosteric activation by a specific mucopolysaccharide" (25). In contrast, Kaufman and co-workers (67) interpreted their own findings [in particular the general activation by polyanions and salts and the screening effect of salts on the activation by polyanions (66, 67)] by proposing that the activation by polyanions does not represent an effect of "specific mucopolysaccharides." Instead they suggest that the effect depends primarily on the negative charge density of the macromolecule (and to some extent on the nature of the cation) and an electrostatic interaction of these molecules with the enzyme. According to this idea, the apparent specificity for certain mucopolysaccharides, for example, the greater activity of heparin as compared to chondroitin sulfate, would simply be a reflection of the greater negative charge density of the former molecule. Indeed, according to this interpretation, it was postulated that there might not be a fundamental difference between the actions of small anions and highly negatively charged polymers (67).

The preceding interpretation of the activation of TH by polyanions led to the formulation of a structural model in which it was postulated that the hydroxylase exists in two or more states that differ in catalytic activity and in either tertiary or quaternary structure, or in the distribution and extent of protonation. In low-salt concentrations and in the absence of polyanions, the form of the enzyme with the higher catalytic activity was assumed to have local areas of higher positive charge density and electrostatic potential than the form of the enzyme with lower activity. According to this model, polyanions would activate TH because they would interact more strongly with the enzyme in its more active state and would stabilize it (67).

Studies carried out with bovine adrenal TH that had been purified after it had been subjected to limited proteolysis by chymotrypsin (M_r = 34,000) indicate that the site on the hydroxylase that is involved in its interaction with polyanions is distinct from the catalytic site (81). It has been found, for example, that whereas the native adrenal enzyme shows the expected stimulation by heparin and phosphatidyl-L-serine, the chymotrypsin-treated enzyme is stimulated by neither of these polyanions. A significant difference between the activation by phosphatidyl serine of the adrenal and brain enzyme is that activation of the former enzyme not only involves a decrease in the K_m for 6MPH$_4$, but also an increase in V_{max} (81).

Not only have many of the observations on the activation of TH by polyanions and the effects of salts (67) been confirmed, but a similar electrostatic model for the interaction of the enzyme with anions has also been put forth (89). In addition, the observation that the stimulating effect of polyanions is lost after the enzyme has been subjected to limited proteolysis (81) has also been replicated (89).

It has already been mentioned that the demonstration that PAH is markedly stimulated by phospholipids (82), when taken together with the notion that the three pterin-dependent aromatic amino acid hydroxylases constitute a family of enzymes with similar properties (90), prompted the studies showing that TH is also activated, albeit by a different mechanism, by phospholipids (66).

Another example of the similarities between the members of this family of enzymes is provided by the observation that rat brain TH is activated by limited proteolysis by trypsin (91). As is the case of activation by phospholipids, the manner in which the activation is expressed is not the same for the two enzymes. Thus, PAH is activated 25–30-fold by either phospholipids or limited proteolysis, and the higher activity is expressed mainly as an increase in V_{max} with a small ($\sim$50%) decrease in the K_m for phenylalanine (in the presence of BH_4), whereas the K_m for BH_4 actually increases (82). In contrast, limited proteolysis of brain TH by trypsin leads to an activation that is more modest and appears to involve mainly a decrease in the K_m for both tyrosine (near fivefold) and $DMPH_4$ ($\sim$ sevenfold). Since these studies were carried out on a crude enzyme preparation, the effects of this treatment on V_{max} could not be determined. In sharp contrast with the ability of heparin to activate native striatal TH (25), this polyanion was found to inhibit brain enzyme that had been activated by limited proteolysis with trypsin (91).

As far as the effects of polyanions are concerned, bovine adrenal TH that has been subjected to limited proteolysis with chymotrypsin (M_r = 34,000) shows both similarities and differences when compared to the effects of polyanions on the trypsin-digested brain enzyme. With the proteolyzed bovine adrenal enzyme, heparin neither stimulates nor inhibits, but phosphatidyl serine markedly inhibits (81). In other words, in this respect the inhibitory effect of phosphatidyl serine on the proteolyzed adrenal enzyme is similar to the inhibitory effect of heparin on the proteolyzed brain enzyme. These results indicate that although heparin and phosphatidyl serine affect the activity of brain TH in very similar ways, their effects on the chymotrypsin-treated bovine adrenal enzyme are different, heparin having little, if any, effect and phosphatidyl serine inhibiting. The finding that heparin stimulates the native bovine hydroxylase activity more at low than at high $6MPH_4$ concentrations indicates that the heparin stimulation involves a decrease in K_m for the pterin and an increase in V_{max} (81). Vigny and Henry (89) also found that heparin decreased the K_m for DMPH$_4$ of the bovine adrenal hydroxylase. These effects of heparin on the K_m of the pterin cofactor are in contrast to the report that heparin did not affect the K_m of the adrenal enzyme for DMPH$_4$ (91).

C. REGULATION BY PHOSPHORYLATION

Although a convincing case cannot yet be made in favor of the idea that activation of TH by limited proteolysis or by polyanions (such as heparin and certain phospholipids) is of physiological significance, these early studies were nevertheless important because they foreshadowed the changes in properties of the enzyme that probably are involved in its physiological regulation. Specifically, results of these studies made it seem likely that regulation of the activity of this enzyme *in vivo* would be manifested by an increase in velocity, especially at physiological pH values (66), a modest increase in the affinity of the enzyme for its pterin cofactor, with a concomitant decrease in affinity for the end products in the pathway for catecholamine biosynthesis, dopamine, and norepinephrine.

Next, mechanisms that might be involved in the acute regulation of the activity of the enzyme, rather than those that lead to its long-term, chronic control will be considered first. Whereas the latter type of modulation, which will be considered in Section VII, probably involves *de novo* synthesis of new hydroxylase molecules (i.e., enzyme induction), synthesis of new enzyme molecules is probably not involved in its acute regulation.

That TH is capable of being acutely regulated has been recognized since the 1950s. In that period, catecholamine synthesis in adrenal and nerve tissue was shown to be stimulated in response to increased nerve activity (92–99). Since stimulation of dopamine- and norepinephrine-containing tissues usually leads to a parallel increase in both synthesis and release of catecholamines, the tight coupling between these two events permits these tissues to maintain steady-state levels of catecholamines. Furthermore, since catecholamines inhibit TH *in vitro* (12, 100), the first mechanism proposed to explain this acute activation of TH assumed that activation was due to a decrease in a small pool of catecholamines that normally inhibit the enzyme (12, 96, 101, 102). This kind of mechanism would not necessarily demand that the hydroxylase isolated from the stimulated tissue be in an activated state, since in this hypothesis, it is the enzyme's intracellular environment, rather than the enzyme itself, that would be altered by the activating stimulus.

Not only did subsequent studies fail to provide support for the above hypothesis (103), but evidence began to accumulate indicating that the hydroxylase itself might be altered in stimulated tissues. Furthermore, the evidence suggested that this alteration might involve phosphorylation of the enzyme.

One of the clearest indications that activation of TH *in vivo* can lead to a rather stable modification of the enzyme as evidenced by an increase in the apparent affinity of the enzyme for its pterin cofactor and a decrease in its affinity for end product inhibitors came from pharmacological studies. Zivkovic et al. (104), showed that injection of neuroleptics such as haloperidol, methiothepin, and reserpine into rats leads to the acute activation (within 1–2 h) of striatal TH, but not of the enzyme in noradrenergic tissues such as hypothalamus. The activation followed the pattern described above, that is, the enzyme from rats treated with neuroleptics had a low apparent K_m for the pterin cofactor (DMPH$_4$), with no change in K_m for tyrosine and no change in V_{max}, and decreased inhibition by dopamine. The alteration in the hydroxylase molecule that leads to these altered kinetic properties was not identified.

It is of interest that, in contrast to the short-term activating effect of reserpine, which appeared to be due to a change in the catalytic properties of existing TH molecules, the long-term (3–4 days) effect of this drug in noradrenergic neurons (but not in dopaminergic neurons), was shown to be due to an increased number of hydroxylase molecules (105). The results of these two studies indicate that reserpine leads to acute activation of the enzyme in dopaminergic neurons, whereas it selectively induces the enzyme in noradrenergic neurons. Taken together, these findings suggest that in the latter type of neuron, under the influence of reserpine, activation of the enzyme is not a necessary step in its induction.

In the same year, Morgenroth et al. (106) reported that TH in guinea pig vas deferentia was activated after electrical stimulation or potassium depolarization of the sympathetic nerves in this tissue. Furthermore, they showed that the enzyme in the high-speed supernatant fraction obtained from the stimulated tissue was still in an activated state (five- to sixfold decrease in K_m for $DMPH_4$ and for tyrosine, decreased affinity for end product inhibitors, and a 50% increase in V_{max}), an indication that in response to stimulation the hydroxylase had undergone a rather stable modification that altered its catalytic properties.

Since it was also known that both electrical stimulation and depolarizing agents can increase the concentrations of cyclic adenosine $3',5'$-phosphate (cAMP) in brain slices (107, 108), it seemed likely that activation of TH in response to nerve stimulation might be mediated by a cAMP-dependent process.

An indication that TH could indeed be activated by cAMP was obtained when it was shown that dibutyryl-cAMP added to rat striatal slices stimulated the conversion of ^{14}C-tyrosine to ^{14}C-dopamine (109). This point was proven directly in experiments with both homogenates and a soluble fraction prepared from rat striatum in which it was shown that added dibutyryl-cAMP stimulated TH and that the stimulation was also associated with a decrease in K_m values for both tyrosine and $DMPH_4$, and an increase in K_i for dopamine (110). That is, added cAMP produced changes in TH that were similar to those seen after nerve stimulation.

That activation of TH by cAMP probably involves phosphorylation of some kind of protein was strongly indicated by the finding that partially purified cAMP-dependent protein kinase (PKA) added to high-speed supernatant fractions from rat hippocampi increased the activity of TH severalfold (111).

Lloyd and Kaufman (57), using a highly purified preparation of TH from bovine caudate, and Lovenberg et al. (112), using a crude rat striatal enzyme fraction, independently demonstrated that this hydroxylase could be activated by PKA. With the purified bovine enzyme, it was shown that the activation was due to a fourfold decrease in the K_m for the pterin cofactor (either BH_4 or $6MPH_4$) (57). A similar decrease in the K_m for $6MPH_4$ was reported with the crude rat brain enzyme (112). These results with BH_4 demonstrated for the first time that activation of the hydroxylase by phosphorylation was associated with a decrease in the K_m of the enzyme for the naturally occurring pterin cofactor. In contrast to the results obtained by Harris et al. (110) with crude rat striatal TH, however, activation of the purified bovine brain enzyme (57) or the crude rat brain enzyme (112) did not lead to a decrease in the K_m for tyrosine.

Significantly, activation of the purified bovine brain enzyme led to a marked shift in the pH optimum from pH 6.0 to 7.4. It was pointed out (57) that all of the changes in kinetic properties of TH after activation by exposure to phosphorylating conditions were essentially the same as those seen when the enzyme was activated by phosphatidyl-L-serine (66). Another indication that these diverse processes lead to similar changes in the enzyme's kinetic properties was the finding that the two types of activation were not additive (57).

Crude bovine adrenal TH was also shown to be activated by exposure to ATP and PKA, with the activation being expressed at pH 6.4 as a decrease in the apparent K_m for $6MPH_4$ and an increase in V_{max} (81). In contrast to the results obtained with purified bovine brain enzyme (57), however, activation of bovine adrenal TH led to only a slight shift in the pH optimum (from pH 6.2 to 6.4). With respect to differences between adrenal and brain TH, it was found that activation of the adrenal enzyme by phosphatidylserine resembles activation by phosphorylation, that is, the K_m for the pterin cofactor is decreased and V_{max} increased (81). These results provide further evidence for the marked similarity between the effects of phosphorylation and phosphatidylserine on TH: with brain TH, both modes of activation are expressed by decreases in the K_m for the pterin cofactor without an increase in V_{max}, whereas with adrenal TH, both of these activations lead to decreases in the K_m for the pterin cofactor and increases in V_{max} (57, 81). Another similarity between these two types of activation is that they are both lost when the bovine adrenal TH is subjected to limited proteolysis (81).

The observation that purified TH could be activated by PKA made it seem likely that this activation was the result of a direct phosphorylation of TH. Despite these expectations, the first attempts to show direct phosphorylation of TH from bovine brain (57) or rat brain (113), were unsuccessful.

Yamauchi and Fujisawa (114) reported that partially purified bovine adrenal TH can be activated in vitro by PKA kinase. They demonstrated that the observed activation was accompanied by the incorporation of an undetermined amount of ^{32}P from (^{32}P)ATP into the TH molecule. Although the effect of phosphorylation on the K_m values for the cofactor and substrate were not determined, the activation ($\sim$ threefold at pH 6.8) was associated with a shift in the pH optimum of the enzyme from 6.0 to 6.8, confirming the original observation of Lloyd and Kaufman (57).

Yamauchi and Fujisawa (115) also showed that TH in bovine adrenal medullary extracts can be activated by an endogenous PKA and that this activation can be reversed by the action of an endogenous phosphatase. These results showed the two enzymes that would be needed for phosphorylation–dephosphorylation of TH (a protein kinase and a phosphatase) are present in bovine adrenal medulla.

Vulliet et al. (116) also reported that TH purified from rat pheochromocytoma is a substrate PKA. Activation of the enzyme, which was expressed as a substantial decrease in the K_m for 6MPH$_4$ (from 480 to 120 μM at pH 6.2) with no change in the K_m for tyrosine or increase in V_{max}, was accompanied by the incorporation of about 0.7 mol of inorganic phosphate (P$_i$) per 60,000 M_r subunit of TH. The amount of protein-bound P$_i$, if any, present in the isolated enzyme prior to exposure to adenosine triphosphate (ATP) and PKA, was not determined. There was a good correlation between the amount of ^{32}P$_i$ incorporated and the degree of activation during the first 15 min of incubation with the phosphorylating system. Beyond that period, TH activity fell and ^{32}P$_i$ incorporation continued, as if the enzyme was being inactivated by further phosphorylation. Vrana et al. (117) reported similar observations, which indicated a phosphorylation-mediated loss of hydroxylase activity.

In this report by Vulliet et al. (116), as well as others that have been published by Weiner and co-workers (118), the double-reciprocal plots of initial velocities versus concentrations of the tetrahydropterin for the nonactivated hydroxylase did not yield straight lines. Rather, at high concentrations of the pterin, there was a pronounced downward deviation because of a greater increase in the velocity at these higher substrate concentrations than would be expected for simple Michaelis–Menten kinetics. Since activation of the enzyme eliminated most, if not all of this deviation, it was assumed that the break in the double-reciprocal plot is due to the contributions of two different forms of the enzyme, a low K_m and a high K_m form, the former corresponding to the phosphorylated and the latter to the non-phosphorylated form of the enzyme. This interpretation of the biphasic double-reciprocal plots is supported by the finding that activation of the hydroxylase by exposure to ATP and PKA eliminates this kinetic anomaly (119). On the other hand, this interpretation must be reconciled with the fact that not all kinetic studies of the hydroxylase have observed biphasic double-reciprocal plots of velocity versus pterin concentration (52, 66, 83, 113, 120). The failure of other workers to consistently observe this kinetic behavior may be due to the dephosphorylation of the phosphorylated form of TH in some preparations of the enzyme, thereby converting a mixture of two forms of the enzyme into only a single one with the elimination of biphasic kinetics. Somewhat at variance with this possibility, most studies in which these funny kinetics were observed were carried out with relatively crude preparations of TH.

Studies on the activation of brain TH by phosphorylation lagged behind those on the enzyme from other tissues such as adrenal medulla or pheochromocytoma. Following the demonstration that incorporation of unknown amounts of $^{32}P_i$ into the enzyme purified from rat caudate nucleus activated the enzyme (121), Edelman et al. (122), showed that PKA catalyzes the incorporation of 0.7–0.9 mol of ^{32}P/mol of purified bovine striatal TH subunit. No attempt, however, was made to correlate the extent of phosphorylation with the extent of activation. Subsequently, activation of the enzyme by PKA was shown to be proportional to the amount of $^{32}P_i$ incorporated up to about 0.8 mol/mol of subunit (123, 124). In the latter study, a significant deviation from a linear relationship was observed with a disproportionately large increase in activity associated with the incorporation of between 0.5 and 0.75 mol $^{32}P_i$/mol of subunit (124).

The results discussed above proved that exposure of TH from different tissues to protein phosphorylating conditions activates the enzyme. Despite this consensus, there was a considerable amount of disagreement about the kinetic expression of this activation. Whereas most results agreed with those first reported by Lloyd and Kaufman (57), which showed that activation of purified TH by PKA is expressed as a decrease in the K_m for the pterin coenzyme, with both the K_m for tyrosine and the V_{max} remaining essentially unchanged, others reported that the activation is expressed exclusively as a change in V_{max} at pH 5.5 (121). Still others, as will be discussed later in this section, claim that the enzyme can only be activated to the most modest extent (12–25% decrease in K_m for 6MPH$_4$, little change in V_{max}) in gel-filtered rat striatal extracts because the filtration step itself activates the enzyme to almost the same extent as that seen after phosphorylation by PKA, with both treatments decreasing the K_m for 6MPH$_4$ (125).

A clue that may explain some of the divergent results in this area was provided by a study of how the K_m for the pterin cofactor and V_{max} vary with pH for both the control and the activated hydroxylase. Pollock et al. (51) found that for partially purified bovine striatal TH, the way in which activation by phosphorylation is expressed is critically dependent on the pH at which the hydroxylase activity is measured. As seen in Figure 2, at pH 6.0 activation is expressed as a sharp decrease in the K_m for the pterin cofactor with little change in V_{max}, whereas at pH 7.0–7.2 activation is expressed as a large increase in V_{max} and a much smaller decrease in K_m.

Similar results were found with gel-filtered rat striatal extracts (51). One difference between the partially purified bovine enzyme (Fig. 2) and this crude rat enzyme (Table VII) is that with the latter preparation the velocity (V_{max}) of the activated enzyme is still higher at pH 6.2 than it is at pH 7.2. It should be noted, however, that even with the rat enzyme the shape of the pH–V_{max} curve has been dramatically altered by activation, so that for the control enzyme V_{max} declines by a factor of 50 on going from pH 6.2 to 7.2, whereas for the activated enzyme V_{max} declines by only a factor of 3.3 for the same shift in pH.

One of the questions that these results helped to clarify is whether activation of the enzyme by exposure to protein phosphorylating conditions leads to a shift in the pH optimum of the enzyme. Lloyd and Kaufman (57) first reported such an alkaline shift for the bovine striatal enzyme, with the pH optimum increasing from 6.0 for the control to pH 7.4 for the activated enzyme. Although these results were qualitatively confirmed by Lazar et al. (52) (optimum pH shifted from 6.0–6.2 to pH 6.8) for bovine striatal TH, an alkaline shift in pH optimum was not reported by Hegstrand et al. (126) or Goldstein et al. (127) for the rat striatal enzyme.

As is evident from the results in Figure 2 and Table VII, the pH profile of the activated enzyme depends on the cofactor concentration used in the assay. For both the bovine and rat striatal hydroxylase, at less than saturating cofactor concentrations, the pH optimum of the activated enzyme would be expected to show a marked alkaline shift of the pH optimum. At saturating concentrations of the

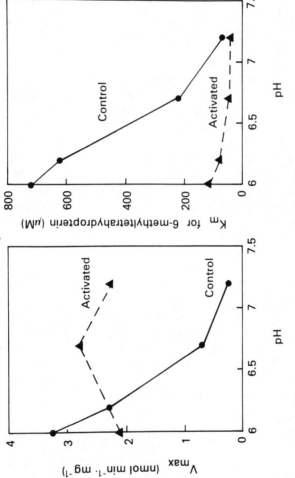

Figure 2. Kinetic parameters of activated and control TH at different pH values. Partially purified bovine striatal TH was incubated at pH 7.0 with beef brain protein kinase without ATP (control) or with ATP (activated). Aliquots from each incubation were then further incubated at the indicated pH with the components of the TH assay plus EDTA to stop further protein kinase activity. Six different concentrations of 6MPH₄ were used. [From Pollock et al. (51).]

TABLE VII
Effect of pH on Kinetic Parameters of Activated and Control Rat Striatal
Tyrosine Hydroxylase[a]

State of the Enzyme	pH	K_m for 6MPH$_4$ (μM)	V_{max} (nmol/min)
Control	6.2	2200	0.100
	6.7	1100	0.017
	7.2	220	0.002
Activated	6.2	840	0.073
	7.2	93	0.022

[a] Data from Pollock et al. (51).

pterin cofactor, the bovine enzyme would still be expected to show the alkaline shift (see Fig. 2), whereas the rat striatal enzyme would not (Table VII). Consistent with this generalization, it has been reported that even for rat striatal TH, assays carried out at subsaturating 6MPH$_4$ concentrations show an alkaline shift of the pH optimum on activation (51).

These results, which were reported by Pollock et al. (51) with purified bovine striatal TH, were confirmed by Lazar et al. (52). Lazar et al. also found that activation of bovine striatal TH by phosphorylation was expressed as a decrease in the K_m for the pterin cofactor at pH 6.0, with little change in V_{max}, whereas at pH 7.0, the K_m for the pterin decreased and V_{max} increased.

Although they did not study the effect of pH on the kinetic constants of control and activated TH, Hegstrand et al. (126) and Goldstein et al. (127) both reported that the activation by PKA of crude rat striatal TH was greater at pH 6.8–7.2 than at pH 5.6–6.0, the latter range being close to the optimum for the control enzyme. Since these workers used nonsaturating concentrations of the pterin cofactor, their "activity" measurements probably reflect decreases in K_m for the cofactor as well as increases in V_{max}. Therefore, their finding that the activation was more pronounced at higher pH values is consistent with, and can be explained by, the results of Pollock et al. (51).

The demonstration that TH in gel-filtered rat striatal extracts can be activated by PKA and that the activation at pH 6.2 is expressed as a decrease in the K_m for 6MPH$_4$ (51), disagrees with the results reported by Ames et al. (125) who, as mentioned earlier, observed only very modest activation of TH in gel-filtered rat striatal extracts. Since their assays were carried out at pH 6, a large decrease in the K_m for the pterin cofactor would have been expected from the results of Pollock et al. (51) and Lazar et al. (52). In agreement with the results of Pollock et al. (51) but in contrast to those of Ames et al. (125), Vrana et al. (117) also demonstrated activation of TH by exposure of gel-filtered rat striatal extracts to phosphorylation conditions. Furthermore, Vrana et al. (117) were unable to replicate the finding of Ames et al. (125) that gel filtration of striatal extracts by itself led to a decrease in K_m for 6MPH$_4$. To explain their failure to observe significant activation of TH in gel-filtered extracts of rat striatum, Ames et al. (125) suggested that activation of TH by phosphorylation is dependent on the presence of catecholamines. Although this explanation does not account for positive results such as those reported by Pollock et al. (51), and by Vrana et al. (117),* recent evidence, to be discussed later, does indicate a connection between catecholamine inhibition of the hydroxylase and its activation by phosphorylation.

The data shown in Figure 2 and Table VII clarify several other confusing issues about regulation of TH activity. First, they indicate that the K_m values for the pterin cofactor that have been determined at the pH optimum of the nonactivated enzyme ($\sim$ pH 6) probably have little, if any, relevance to the physiological regulation of the enzyme. These values, as already noted, appeared to be alarmingly high when compared to tissue levels of BH$_4$. This apparent disparity

---

* The explanation for the glaring discrepancy between the results of Pollock et al. (51) and Vrana et al. (117) on the one hand and those of Ames et al. (125), on the other hand, is not known. A possible explanation for part of the discrepancy can be suggested from a methodological difference between the two sets of results. In the former two studies, where activation of the enzyme by PKA was observed in gel-filtered striatal rat extracts, fresh tissue was used. Ames et al. (125), however, used frozen tissue. It is possible that the process of freezing and thawing releases inhibitory catecholamines from organelles and that their removal by gel filtration of extracts activates the hydroxylase. In contrast, the concentration of catecholamines in extracts of fresh unfrozen striatal tissue may be sufficiently low that gel filtration would have no effect. It is of interest, in this regard, that Okuno and Fujisawa (191) reported that TH in crude extracts prepared from frozen rat striata was almost inactive at pH 7 but that acid precipitation of the enzyme led to activation, presumably by removal of inhibitory dopamine.

generated a lot of needless hand-wringing about how severely limited TH might be by the availability of its pterin coenzyme. The finding that the K_m for the pterin cofactor is at least an order of magnitude less at pH 7.2–7.4 than it is at pH 6 (Fig. 2, Table VII), indicates that the problem of cofactor available has probably been grossly overstated. Coherent with this conclusion that TH does not normally operate under such a bizarre handicap, it has been found that in rat striatal synaptosomes, at physiological pH, the enzyme appears to be operating with endogenous pterin cofactor concentrations that are much closer to saturation than would have been predicted from results of cell-free assays carried out at pH 6.0–6.2 (51). Thus, it was found that the addition of saturating amounts of BH_4 to synaptosomes increased the *in situ* activity of TH by only 50%, a result that would be difficult to reconcile with the very high K_m values for BH_4 that are found at pH 6.0–6.2, but which is readily understood by the lower K_m value.

The results of Pollock et al. (51) (Fig. 2) also predict that in the physiological pH range, activation of TH by PKA would be expressed primarily as an increase in V_{max}. Based on these *in vitro* results, it can be anticipated that in an intact tissue preparation maintained at physiological pH, the extent of activation of TH by PKA would be relatively independent of the BH_4 concentration. To test this prediction, the degree of activation of the enzyme in rat striatal synaptosomes by exogenous 8-bromo-cAMP was studied as a function of varying concentrations of BH_4 added to the medium. In accord with the above expectations, it was found that the percent stimulation of *in situ* TH activity by the cAMP derivative was relatively independent of added BH_4 concentrations; at 1 mM BH_4 in the medium, for example, 8-bromo-cAMP stimulated to about the same extent in the absence (72% stimulation) and in the presence (79% stimulation) of 1 mM BH_4, a result that is consistent with the conclusion that the cAMP derivative was stimulating TH mainly by an increase in V_{max}. Also supporting this conclusion was the finding that the added cyclic nucleotide led to only a slight decrease in the apparent K_m for the exogenous BH_4 (0.34 vs. 0.27 mM) (51).

Most of the early work on the *in vitro* activation of TH by phosphorylation focused on the activation mediated by PKA. Results of these studies appeared to have obvious relevance to the physiological regulation of the enzyme, particularly to the way the hydroxylase is activated in response to neuronal stimulation. As already discussed briefly, the changes in the kinetic properties of the neuronal enzyme after *in situ* activation by acute electrical nerve stimulation or chemical depolarization resemble those of the isolated enzyme after activation by PKA-mediated phosphorylation. Thus, although Cloutier and Weiner (103) first reported that stimulation of the hypogastric nerve of the vas deferens preparation of the guinea pig leads to activation of TH that is expressed in a way that is not the same as that seen after PKA activation *in vitro*, (i.e., a twofold increase in V_{max} with no change in the K_m for DMPH$_4$), subsequently, the opposite change, that is, at pH 6, a decrease in the K_m for 6MPH$_4$ with no change in V_{max} (128) was reported. The decrease in K_m for the pterin cofactor was in accord with earlier results reported by Morgenroth et al. (106) who, in addition to a decrease in the K_m for the pterin cofactor, also observed an increase in V_{max} (measured at pH 7.4) and a decrease in the K_m for tyrosine. Although the increase in V_{max} and decrease in K_m for the pterin are coherent with the expected effect of PKA on the hydroxylase at this pH (see Fig. 2), the reported decrease in the K_m for tyrosine has not been replicated by others. It is of interest, however, that Harris et al. (110) also reported a similar exceptional finding with TH that had been activated *in vitro* by treatment with cAMP. Additional strong support for the idea that activation of the hydroxylase by nerve stimulation is mediated by PKA, came from the finding that, just like activation by this kinase *in vitro*, the activation *in situ* was also characterized by an increase in K_i for the feedback inhibitor dopamine (129).

Further substantial support for the idea that activation of TH in neuronal tissue in response to nerve stimulation is mediated by PKA (110, 130) came from the finding that the addition of cAMP to the neuronal preparation appeared to activate the enzyme in a way that was indistinguishable from that seen after activation by electrical stimulation of the nerve (129). Perhaps the most powerful argument of all for the PKA thesis was the observation that added cAMP could not further activate the enzyme after it had been activated by nerve stimulation (129). Finally, the report that both electrical stimulation and depolarizing agents increase the concentration of cAMP in brain slices (107, 108), as well as the finding that isobutyl-methylxanthine, an inhibitor of phosphodiesterase that would decrease the degradation of cAMP, enhanced the activation of TH resulting from nerve stimulation (130), seemed to provide the clincher.

Implicit in the notion that depolarization-mediated activation of TH is catalyzed by PKA is a "one-site, one-kinase" model for the physiological regulation of the enzyme. An added attractive feature of this model was the fact that it is analogous to the one that had been proposed to account for the activation of PAH by PKA, an activation that results from the phosphorylation of a single site (131, 132).

Despite what seemed to be an unassailable case in favor of it, there were many signs that this "one-site, one-kinase" model for TH, where the kinase is PKA, was oversimplified. An early cautionary note was struck by the observation that, unlike brain slices, depolarization of synaptosomes by treatment with either high K^+ or veratridine does not lead to elevation of cAMP levels (133), a finding that is coherent with other evidence that the accumulation of cAMP in brain slices following neuronal depolarization most likely occurs in postsynaptic cells (134), whereas TH is located mainly presynaptically. Another serious blow to the case came from the failure to replicate what appeared, at the time, to be one of its most powerful supporting arguments, that is, the finding that cAMP added to unstimulated neuronal tissue mimicked the effect of nerve stimulation, but was without effect when added to the stimulated tissue. Thus, whereas Weiner et al. (130), using the guinea pig vas deferentia preparation, also found that cAMP added to the unstimulated control mimicked the effect of nerve stimulation, they found, in contrast to the result of Murrin et al. (129), that cAMP also activated the hydroxylase when it was added to the stimulated side. Similar results were reported by Goldstein et al. (127) with synaptosomes treated with the depolarizing agent veratridine, that is, the activation by cAMP was additive with that produced by depolarization. In time, one of the original proponents of the notion that PKA is responsible for the activation of TH induced by depolarization reported that dibutyryl-cAMP still activated the enzyme even when added to striatal slices that had been exposed to depolarizing concentrations of potassium (135). These discordant findings could not be easily reconciled with the notion that PKA was the mediator of the *in situ* activation of TH in response to neuronal depolarization.

Even more difficult to ignore or accommodate was the compelling evidence that Ca^{2+}, whose entry into neuronal cells is stimulated by electrical activity (136) or exposure to depolarizing agents, such as high K^+ or veratridine (133), is essential for the depolarization-induced activation of TH but not for the cAMP-induced activation (106, 137–139).

The evidence that has been discussed undermined the "one-site, one-kinase" model for the physiological regulation of TH. Moreover, the results raised the possibility that one or more Ca^{2+}-dependent protein kinases might play a role in controlling the activity of the enzyme, especially the activation seen in response to increased neural activity and chemical depolarization.

The first conclusive evidence that a Ca^{2+}-dependent protein kinase could activate TH was reported by Yamauchi and Fujisawa (140) who found that the hydroxylase in rat brainstem cytosol can, in the presence of ATP and Mg^{2+}, be activated not only by cAMP but also by Ca^{2+}, with the Ca^{2+} activation showing an absolute requirement for calmodulin. The additional finding that the cAMP activation and the Ca^{2+}-calmodulin activation were additive suggested different mechanisms for these two modes of activation.

When the Ca^{2+}-calmodulin activation was studied with purified bovine-adrenal TH, it was found that the enzyme could be phosphorylated by a Ca^{2+}–calmodulin-dependent kinase but that the phosphorylated enzyme had not been activated (141). Addition of another component, designated "activator protein," was required to achieve activation of the phosphorylated enzyme. The maximum activation was about twofold and, in contrast to activation by PKA, was expressed exclusively as an increase in V_{max} (142).* These findings indicated that activation and phosphorylation by Ca^{2+}–calmodulin-dependent protein kinase are two distinct reactions: The first one involves phosphorylation of the hydroxylase by the kinase and the second involves activation of the phosphorylated hydroxylase by

* Unfortunately, there is some confusion about the catalytic properties of TH that has been phosphorylated by Ca^{2+}–calmodulin-dependent protein kinase. The problem arose when it was reported, without any data presentation, that activation is expressed as a decrease in the K_m for 6MPH$_4$ (140). Although this error was subsequently corrected, with the statement that "activation by calmodulin-dependent protein kinase II was caused exclusively by an increase in V_{max} (151)," the erroneous statement was disseminated in a recent review (196).

the activator protein by a mechanism that has yet to be explored. The activator protein was not required for the Ca^{2+}–calmodulin-dependent phosphorylation of TH, nor did it have any effect on the activity of the unphosphorylated enzyme or, it should be emphasized, on the activity of the enzyme phosphorylated and activated by the PKA.

The purified activator protein has a molecular weight of 70,000 (143) and is a dimer of identical subunits, $M_r = 35,000$. Unlike Ca^{2+}–calmodulin-dependent protein kinase, which has been reported to be present only in brain, particularly in cerebral cortex, brainstem, and cerebellum (144), the activator protein appears to be ubiquitous. It is present in all of the subcellular fractions, although the highest relative specific activity was found in the cytosol. In addition to brain, it was found in a number of peripheral tissues, such as adrenal glands, liver, heart, and skeletal muscle; however, its concentration in these tissues is lower than that in brain.

The presence of activator protein in tissues that contain neither TH nor TPH suggested other roles for this protein. This notion has been reinforced by the finding that a soluble acid protein, designated the 14-3-3 protein by Moore and Perez (145), is present in high concentrations in brain tissue (where it is localized in neurons) is identical to the activator protein (146). Its tissue distribution correlates closer to that of calcium-dependent protein kinases than to the aromatic amino acid hydroxylases, a pattern indicating that the 14-3-3 protein is involved in other metabolic processes that are modulated by these kinases. Also coherent with this view is the finding that there is strong homology between the 14-3-3 protein and an endogenous inhibitor of Ca^{2+}–phospholipid-dependent protein kinase (PKC) (147). This inhibition by the 14-3-3 protein raises the possibility of a reciprocal relationship between the effects of PKC and CaM–PKII on TH.

The ability of calcium to activate TH in slices or in the soluble fraction of a tissue homogenate varies from tissue to tissue and also within the central nervous system. As mentioned previously, TH in the cytosol from adrenal medulla can be activated by a cAMP-dependent phosphorylation and, moreover, ^{32}P from $(-^{32}P)ATP$ is incorporated into the enzyme (114). In contrast to the brainstem preparations, however, addition of Ca^{2+} to the cytosol from adrenal medulla did not lead to either activation or to the incorporation of

^{32}P (115). The reason for this lack of effect of Ca^{2+} has been traced to the fact that although adrenal medulla has both calmodulin and activator protein, it lacks Ca^{2+}–calmodulin-dependent protein kinase (143). Purified TH from adrenal medulla can be activated by incubation with Ca^{2+}, calmodulin, Ca^{2+}–calmodulin-dependent protein kinase, activator protein, ATP, and Mg^{2+}.

Added Ca^{2+} does not activate TH in various preparations of rat brain striatum (148, 149). Kapatos and Zigmond (150) failed to find any stimulation by calcium of TH activity in synaptosomes or the soluble fraction prepared from rat brain striatum. On the other hand, TH purified from rat striatum is activated when it is incubated with Ca^{2+}–calmodulin protein kinase and the activator protein (142).

Morgenroth et al. (148) reported an activation of striatal TH by EGTA (ethylene glycol-bis (β-aminoethyl ether)N,N,N′,N′-tetra-acetic acid) whereas Lerner et al. (149) reported that neither Ca^{2+} nor EGTA had any effect on the striatal hydroxylase. The lack of effect of calcium on the activity of TH in rat brain striatum, an area of the brain with dopaminergic innervation, is in sharp contrast to the marked effect of calcium in certain other areas rich in noradrenergic nerve endings, namely, the medulla and the pons (148). In all likelihood, the inability of added Ca^{2+} to activate TH in striatum is because this area of the brain, like adrenal tissue, does not have any Ca^{2+}–calmodulin-dependent protein kinase.

The calmodulin–Ca^{2+}-dependent protein kinase that catalyzes the phosphorylation of TH and, also TPH, has been designated CaM–PKII (151). It occurs only in brain tissues, has a molecular weight of 540,000, and is composed of subunits, $M_r = 55,000$ (151). Fujisawa and co-workers (151) also demonstrated that rat brain cytosol contains two other Ca^{2+}–calmodulin-dependent protein kinases with substrate specificities that are distinct from that of CaM–PKII. They have designed these other enzymes as kinase I, molecular weight about 1,000,000, which appears to be similar to brain phosphorylase kinase, and kinase III, molecular weight about 100,000, which appears to be similar to myosin light chain kinase.

Although CaM–PKII has been reported to be present only in brain (144), a very similar enzyme, referred to as the "calmodulin-dependent multiprotein kinase," is present in skeletal muscle (152) and liver (153, 154). Interestingly, this calmodulin-dependent kinase appears to be present in pheochromocytoma cells, raising the possibility that it is also present in adrenal medulla (155). The kinase from skeletal muscle has been shown to be capable of catalyzing the phosphorylation of TH purified from pheochromocytoma at a different site from that phosphorylated by PKA. Phosphorylation of the hydroxylase by the calmodulin-dependent multiprotein kinase did not activate the enzyme (155), a finding that is not surprising in view of the report that the hydroxylase that has been phosphorylated by calmodulin-dependent protein kinase (CaM–PKII) requires activator protein to express its activated state (143). Vulliet et al. (155) did not test the effect of activator protein on the hydroxylase that had been phosphorylated by calmodulin-dependent multiprotein kinase.

Another Ca^{2+}-dependent protein kinase that can catalyze the phosphorylation of TH is protein kinase C (PKC), the Ca^{2+}-activated, phospholipid-dependent enzyme that was isolated and characterized by Nishizuka and co-workers (156).

The first indication that TH might be a substrate for this kinase came from a study by Raese et al. (157). The incorporation into the hydroxylase of an undetermined amount of $^{32}P_i$ from ^{32}P ATP in the presence of this kinase led to activation of the enzyme that was expressed exclusively as a decrease in the K_m for BH_4 (from 1.0 to 0.22 mM), the same change as that seen when the hydroxylase is activated by PKA at pH 6 (see Fig. 2) (51).

Subsequently, it was shown that pure PKC from bovine cerebral cortex can, in fact, phosphorylate and activate TH that has been purified from the cytosol of PC12 pheochromocytoma cells (158). Activation, characterized by a decrease in K_m for $6MPH_4$, no change in K_m for tyrosine or in V_{max}, all measured at pH 7.0, was accompanied by incorporation of about 0.5 mol $^{32}P_i$/mol of M_r = 62,000–60,000 doublet.

In addition to the ability of PKA, CaM–PKII, and Ca^{2+}–phospholipid-dependent PKC to catalyze the phosphorylation of TH, there may be still another one that can carry out this reaction. Andrews et al. (159) reported that an unfractionated extract of rat striatal tissue contains an enzyme that, in the presence of ATP and Mg^{2+}, can activate TH in these tissue extracts, presumably by a process that involves phosphorylation of the enzyme. That the activation is not mediated by PKA was supported by the finding that this activity was insensitive to an inhibitor of this enzyme. Furthermore, activation of the hydroxylase, which was expressed as a twofold increase in V_{max} with no change in K_m for either 6MPH$_4$ or tyrosine, was not affected by the addition of either Ca^{2+} or EGTA, suggesting that the responsible enzyme may not be dependent on Ca^{2+}. It is note-worthy, however, that the change in the catalytic behavior of the hydroxylase is similar to that produced by the action of CaM–PKII. Finally, cGMP-dependent protein kinase phosphorylates and acti-vates TH at pH 6. Activation is expressed just like activation by PKA, that is, at pH 6, K_m for 6MPH$_4$ is decreased with no change in V_{max} (160). Protein kinases that have been tested with highly puri-fied TH (from rat pheochromocytoma) and found to be essentially inactive are phosphorylase kinase, glycogen-synthase kinases 3 and 4, and casein kinases I and II (155).

Before considering the protein kinase(s) that might be involved in the acute regulation of TH in response to nerve stimulation or membrane depolarization, it might be useful to summarize the man-ner in which activation of the hydroxylase by the action of different kinases is expressed. The results shown in Table VIII indicate that the effects of the different kinases on the properties of the hydroxyl-ase are distinct, with activation by CaM–PKII leading to an increase in V_{max} with no change in K_m for the pterin cofactor, activation by PKC leading to the opposite change, and the effects of PKA depend-ing sharply on the pH of the hydroxylase assay.

A knowledge of how the activity of TH is affected by phosphoryla-

TABLE VIII
Changes in the Catalytic Properties of Tyrosine Hydroxylase after Activation[a,b]

Protein Kinase	K_m for Pterin Cofactor	V_{max}
cAMP-dependent (PKA) (pH 6.0)	↑ ↑	0
cAMP-dependent (PKA) (pH 7.0–7.4)	↓	↑ ↑
cGMP-dependent (pH 6.0)	↓ ↓	↑
Ca^{2+}–calmodulin-dependent (CaM–PKII)	0	↑
Ca^{2+}–calmodulin-dependent multiprotein	ND	ND
Ca^{2+}–phospholipid-dependent (PKC) (pH 7.0)	↓ ↓	0
Non-cAMP-dependent (unstimulated by Ca^{2+})	0	↑

[a] The data for cAMP-dependent protein kinase is from Pollock et al. (51). The non-cAMP-dependent kinase that is not stimulated by added Ca^{2+} is that described by Andrews et al. (159); see text for other references.

[b] Not determined = ND; little or no change = 0.

tion by different protein kinases, summarized in Table VIII, is one of the logical prerequisites for a consideration of the mechanism of the acute physiological regulation of the enzyme. Another essential requirement for this analysis is information about what happens to the hydroxylase molecule itself during its *in situ* activation.

An important step toward this goal, and one that moved the analysis beyond a one-step one-kinase model for the regulation of TH toward a more comprehensive picture, was taken by Haycock et al. (161), who studied the *in situ* activation of the enzyme in bovine adrenal medullary cells. In the presence of Ca^{2+}, these cells respond rapidly to treatment with the natural secretagogue acetylcholine, by secreting norepinephrine and epinephrine (162), which, in turn, leads to a compensatory acceleration of catecholamine synthesis due to a Ca^{2+}-dependent, phosphorylation-mediated activation of TH (163). This coupling of accelerated release and synthesis of catecholamines maintains intracellular stores of catecholamines.

Haycock et al. (161) found that ^{32}P-labeled TH isolated from chromaffin cells, when subjected to limited proteolysis with trypsin, yielded two different ^{32}P-labeled peptides that were separated by two-dimensional electrophoresis/chromatography. A differential pattern of labeling of these peptides in response to exogenous 8-bromo-cAMP and acetylcholine indicated that the effects of these two mediators on phosphorylation of TH are carried out by different protein kinases, that is, these results indicated that acetylcholine activated a kinase distinct from PKA.

Two potential candidates for the kinase that might mediate the acetylcholine effect in chromaffin cells are PKC (156) and CaM–PKII (164). The former enzyme is present in adrenal tissue (164) and has been reported to be capable of activating and phosphorylating TH (157). The CaM–PKII, however, has been reported to be absent from adrenal tissue (144). Although these results indicated that of the two kinases that could potentially mediate the acetylcholine activation of TH in chromaffin cells, PKC is the more likely one, it may be recalled that another Ca^{2+}-dependent enzyme, Ca^{2+}–calmodulin-dependent multiprotein kinase is present in PC12 cells, and that it can phosphorylate TH (155).

It may also be premature to dismiss the possibility that PKA mediates the acetylcholine activation of TH. For this possibility to be kept open, one must assume that in these experiments the exogenous

8-bromo-cAMP did not reach the same compartment as the one in which the presumed acetylcholine-induced rise in cAMP takes place. In this regard, note that although it was stated that acetylcholine did not increase cAMP levels in the chromaffin cells (163), it has been reported that the administration of carbamylcholine to rats leads to a prompt 10-fold increase in the content of cAMP in the adrenal medulla (165).

The evidence that in bovine adrenal chromaffin cells, TH can be phosphorylated at different sites by different protein kinases (161) hinted at just how complex the physiological regulation of this hydroxylase might be. Subsequent studies led to the elaboration of a fairly detailed picture of the *in situ* short-term control of TH activity.

Many of these studies were carried out on PC12 cells, a clonal cell line derived from rat pheochromocytoma, a tumor of the sympathetic nervous system that is found most often in the adrenal medulla. The PC12 cells have been utilized extensively as a model system to study many aspects of neuronal physiology, including the mechanism of action of nerve growth factor (NGF), a protein growth hormone required for the *in vivo* and *in vitro* survival and differentiation of sympathetic and sensory neurons (166). Of particular relevance to the present discussion, NGF is known to enhance the phosphorylation of TH in PC12 cells (167).

In addition to NGF, another growth hormone, epidermal growth factor (EGF), which despite its name, stimulates the growth of a variety of cell types, dibutyryl-cAMP, cholera toxin, and phorbol-12-myristate-13-acetate (PMA) lead to rapid phosphorylation of TH in PC12 cells (168). Moreover, potassium depolarization of these cells, which increases intracellular levels of Ca^{2+} (139), also results in activation and phosphorylation of the enzyme (139, 168).

Peptide mapping revealed that four different serine residues in the hydroxylase can be phosphorylated under different conditions. Agents that raise the intracellular level of cAMP (e.g., dibutyryl-cAMP and cholera toxin), and NGF both increase the labeling of two peptides, but the pattern of labeling is not the same. Epidermal growth factor also increased the phosphorylation of one of these two peptides, but also stimulated the phosphorylation of a unique site. PMA, which activates PKC *in situ* (169), increased the phosphorylation of only a single site, the same one affected by EGF and NGF. Potassium depolarization in the presence of Ca^{2+} also increased the phosphorylation of a single unique site, whose phosphorylation was not affected by any of the other mediators. Finally, each of the above agents led to a two- to fivefold activation of TH *in situ* (168). The finding in this study that K^+ depolarization increases the phosphorylation of a unique site is coherent with the report that NGF and K^+ depolarization increase the phosphorylation of different sites (170).

Halegoua and co-workers (171) also found that in PC12 cells, NGF increased the phosphorylation of two peptides; this hormone only enhanced the phosphorylation of one of them in a PC12 mutant deficient in PKA activity. They concluded from these results that NGF mediates the activation of both PKA and PKC and that each kinase can act independently to phosphorylate TH at different sites (171).

Of the two major Ca^{2+}-dependent protein kinases, PKC and CaM–PKII, both of which are present in PC12 cells together with PKA, evidence has been presented indicating that it is the calmodulin-dependent enzyme that mediates the phosphorylation of TH that is enhanced by K^+ depolarization of the cells (172).

Although the results of most studies agree that TH can be phosphorylated *in situ* at multiple sites, there is no consensus about either the number of sites that are phosphorylated or the pattern of phosphorylation observed with different mediators. Thus, the number of sites has varied from two (173) to seven, the latter number being observed in both PC12 cells (174) and in rat superior cervical ganglion (175).

The results discussed above established that TH *in situ* can be phosphorylated at different sites and that the pattern of phosphorylation can vary with different mediators. They did not, however, locate within the peptide chain the sites that are phosphorylated.

This next step in the analysis of the regulation of TH activity by phosphorylation became feasible when the complete amino acid sequence of rat TH was determined (84). This sequence, together with results of studies of the phosphorylation of the pure enzyme by different protein kinases, provided a standard against which results of *in situ* studies of the phosphorylation-mediated activation of TH could be compared.

From an early attempt at this kind of analysis, Vulliet et al. (155) concluded that TH from rat pheochromocytoma is phosphorylated at different sites by PKA and by CaM–PKII. Subsequently, however, it was reported that TH purified from this source is phosphorylated at an identical serine site by PKA, PKC, and CaM–PKII, while the last enzyme also phosphorylates another unique site. In addition, a protein kinase present as a contaminant in the preparation of the hydroxylase used in these studies was found to phosphorylate still another unique site (176, 177). These results obtained with PKA and PKC were in accord with earlier ones published by Albert et al. (158), who reported not only that these two kinases phosphorylated TH at the same serine, but also, as might be expected from this result, that the activation of the hydroxylase produced by these two kinases was the same.

All of the phosphorylation sites mentioned above were localized toward the N-terminus of the hydroxylase, the region believed to represent the regulatory domain(s) of this enzyme, as well as of PAH. The common site phosphorylated by PKA, PKC, and CaM–PKII was identified as Ser[40] (consensus sequence, Arg-Arg-Y-Ser, actual sequence, Arg-Arg-Gln-Ser). The additional unique site phosphorylated by CaM–PKII was shown to be Ser[19] (consensus sequence, Arg-X-Y-Ser, actual sequence, Arg-Ala-Val-Ser), and the one phosphorylated by the contaminating kinase was located at Ser[8] (177). Subsequently, the kinase responsible for phosphorylating this serine residue was identified as a novel proline-directed kinase (whose minimal recognition sequence is X-Ser/Threo-Pro-X (178); the actual sequence in TH is Pro-Ser-Pro-Gln (177). A secondary phosphorylation site for PKA was Ser[153] but this residue was only phosphorylated in a partially proteolyzed preparation of the hydroxylase that had lost about 20 amino acids from the N-terminus, and is therefore not physiologically relevant. The serine residues in TH that are phosphorylated by different kinases are depicted in Figure 3, which also shows some structural details that will be discussed in Section X.

It should be noted that localization of the phosphorylation sites in TH to the NH$_2$-terminal region provides another example of the structural similarity of this enzyme and PAH, since with the latter enzyme the single site that is phosphorylated by PKA with resultant activation (131, 132) is also located in the NH$_2$ terminal region, that is, at Ser[16] (179, 180).

Haycock (181) identified the serine residues in TH that are phosphorylated in PC12 cells in response to a variety of different mediators. Trypsin digestion of ^{32}P-labeled enzyme isolated from the cells

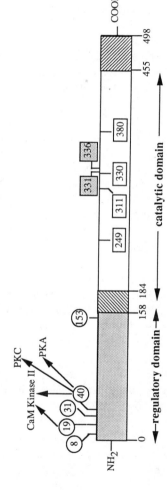

Figure 3. Model depicting the regulatory and catalytic domains of rat pheochromocytoma tyrosine hydroxylase (196, 303, 305). The circles indicate the positions of the phosphorylation sites (Ser[8], Ser[19], Ser[31], Ser[40], and Ser[153]) with their corresponding protein kinases; the rectangles depict the position of conserved cysteines Cys[249], Cys[311], Cys[330], and Cys[380]) and the shaded squares the position of conserved histidines (His[331], His[336]).

produced five distinct labeled peptides containing four acceptor sites: Ser^8, Ser^{19}, Ser^{31}, and Ser^{40}.

Brief (30–60 s) exposure to depolarizing agents such as high K^+, veratridine, and nicotine increased the phosphorylation of Ser^{19}. Pretreatment with the Ca^{2+} chelator EGTA blocked these effects, whereas exposure to A23187, which causes Ca^{2+} influx that is not mediated by membrane depolarization, not only increased the phosphorylation of Ser^{19} but also Ser^{31}.

The growth factor NGF also enhanced the phosphorylation of Ser^{31}. Epidermal growth factor, on the other hand, in sharp contrast with the results of McTigue et al. (168), did not stimulate the phosphorylation of TH in this study.

Agents that increase the intracellular concentration of cAMP, such as forskolin, dibutyryl-cAMP, and 8-bromo-cAMP increased the phosphorylation of Ser^{40}.

Phorbol esters, known to activate PKC, substantially increased the phosphorylation of Ser^{31} and only modestly increased the phosphorylation of Ser^{40}.

Finally, okadaic acid, an inhibitor of phosphatases 1 and 2A, enhanced the phosphorylation of all four of these serine residues, including Ser^8. Treatment with this inhibitor was the only condition that increased phosphorylation of this site. The role of phosphorylation of Ser^8 in the regulation of TH is obscure. With the exception of Ser^{31}, the effect of this inhibitor on phosphorylation of the other sites was considerably greater than that of any other agent.

An analysis of the serine residues in TH that are phosphorylated under various conditions has been extended to dopamine nerve terminals both *in vivo* and in striatal synaptosomes (182). After treatment of control rats with $^{32}P_i$, four sites were labeled, Ser^{31}, Ser^{19}, Ser^{40}, Ser^8, with the last site being labeled the least. Electrical stimulation of the brain area containing the afferent dopaminergic fibers increased the phosphorylation of three out of four of these sites: Ser^{31}, Ser^{19}, Ser^{40}. Brief (5–30 s) depolarization of synaptosomes with elevated K^+ or by treatment with veratridine selectively increased the phosphorylation of Ser^{19}; longer (120–240 s) exposure to high K^+ also labeled Ser^{31}. The PKC activator, phorbol 12,13-dibutyrate, selectively increased $^{32}P_i$ incorporation into Ser^{31} and agents that increase cAMP levels (e.g., forskolin) increased labeling of Ser^{40}. In contrast with the results observed with TH in PC12 cells, treatment with NGF did not stimulate the phosphorylation of the enzyme in synaptosomes. Just as with PC12 cells (181), the only treatment that stimulated the phosphorylation of Ser[8] was exposure to the phosphatase inhibitor, okadaic acid, a treatment that also increased the phosphorylation of the other three sites.

A comparison of the phosphorylation sites *in situ* to those phosphorylated *in vitro* with pure TH reveals one striking difference, that is, *in vitro* Ser[31] is not phosphorylated by any of the known kinases (177), whereas it is phosphorylated in PC12 cells in response to NGF and phorbol esters (181). The phosphorylation of Ser[19] and Ser[31] after exposure of PC12 cells to depolarizing conditions and NGF, respectively, has also been reported by Mitchell et al. (183). In contrast to the results of Haycock (181), however, these workers found that Ser[31] is also phosphorylated following depolarization with high concentrations of potassium.

Cremins et al. (171), based partly on their demonstration that the drugs chlorpromazine and trifluoperazine, known to inhibit PKC *in vitro* and *in vivo,* blocked the phosphorylation of TH induced by NGF and phorbol esters, argued that this kinase is responsible for the NGF- and phorbol ester-mediated phosphorylation of TH at one peptide, T3 (Ser[31]). On the other hand, Haycock (181) concluded that the effects of phorbol esters on TH phosphorylation *in situ* are probably not mediated directly by PKC, but rather by some yet-to-be defined indirect effect of this kinase.

The mystery over the identity of the kinase responsible for the *in situ* phosphorylation of Ser[31] has been at least partially solved with the demonstration that ERK1 and ERK2, two myelin basic protein and microtubule-associated protein kinases, may be involved (184). These enzymes belong to a family of serine/threonine protein kinases whose activity is regulated by the phosphorylation of tyrosine and threonine residues in the kinase (185, 186). The ERKs were identified as the enzymes mediating phosphorylation of TH at Ser[31] (up to 0.6 mol of P_i/mol of TH subunit) in NGF- or bradykinin-treated PC12 rat pheochromocytoma cells. The ERK1-catalyzed incorporation of about 0.2 mol of P_i/TH subunit led to an extremely modest (20–40%) increase in TH activity that appears from the published data not to lead to a decreased inhibition by dopamine and to be associated with neither a decrease in K_m for BH_4, nor an alkaline shift in the pH optimum for the hydroxylase. Just how small the ERK-mediated activation of TH is can be judged by the finding that under the same conditions, incorporation of about 0.3 mol of P_i/TH subunit mediated by PKA activated the hydroxylase by about 15-fold at subsaturating BH_4 concentrations. As the authors themselves acknowledge, it is difficult to assess the physiological significance of this small activation (184).

A summary of the *in vitro* and *in situ* effects of various kinases and agents on the multisite phosphorylation of TH is shown in Table IX.

In a discussion of the functional consequences of phosphorylation of Ser[19] by depolarization, which in all probability is mediated by CaM–PKII and would therefore be expressed as an increase in V_{max} (see Table VIII), Haycock (181) stated that the resulting activation

TABLE IX
Phosphorylation Sites in Tyrosine Hydroxylase[a]

Residue	*In vitro* Kinase	*In situ* Mediators
Serine 8	PDPK[b]	Unknown[c]
19	CaM–PKII	K^+ depolarization, veratridine, nicotine (increase intracellar Ca^{2+}), EGF,[d] A23187
40	PKA PKC CaM–PKII[e] cGMP–PK[f]	Agents that increase cAMP (Forskolin, cholera toxin), phorbol esters,[e] NGF[e]
31	ERK[g]	NGF, EGF[d], phorbol esters, A23187, K^+ depolarization[h]

[a] Data primarily from Campbell et al. (177), Roskowski et al. (160), Vulliet et al. (178), McTigue et al. (168), and Haycock (181).

[b] PDPK, proline-directed protein kinase.

[c] Phosphorylation in PC12 cells is only enhanced by the phosphatase inhibitor, okadaic acid.

[d] Found by McTigue et al. (168), but not by Haycock (181).

[e] Secondary site.

[f] cGMP-dependent protein kinase.

[g] Haycock et al. (184).

[h] Found by Mitchell et al. (183), but not by Haycock (181). Haycock and Haycock (182), however, did find that longer exposure to high K^+ did lead to phosphorylation of Ser[31].

of TH after brief periods of polarization would occur irrespective of cofactor levels. This statement must be qualified. Clearly, no activation would be expressed if cofactor levels were very low. Even at normal levels, the degree of activation that could be achieved would probably be quite modest. This would be true because the K_m for BH_4 of the hydroxylase phosphorylated by CaM–PKII on Ser[19] would, like that of the unphosphorylated enzyme, be much greater than tissue levels of BH_4. This relationship and the fact that activation by this kinase increases V_{max} of TH by less than twofold (142), indicates that the amount of hydroxylase activity that would be expressed after depolarization-induced phosphorylation of Ser[19] would be only a small fraction of the activity that would be expressed after a phosphorylation of Ser[40]. Indeed, if the K_m for BH_4 were as high as 2.3 mM, as claimed by Miller and Lovenberg (58), a value that is 20 times higher than the likely neuronal levels of BH_4 (187), then the amount of activation of TH that could be expected after brief polarization, rather than occurring irrespective of cofactor levels, might be so limited by the low tissue levels of BH_4 as to be physiologically inconsequential and perhaps undetectable.

The potential problem posed by these considerations raises the question of whether there is not also some decrease in the K_m for BH_4 in response to depolarization. Moreover, based on the finding that Ser[31] is also phosphorylated under these conditions (183), it would be a reasonable expectation that phosphorylation of this site would lead to a decrease in the K_m of BH_4. Indeed, if depolarization only results in the phosphorylation of Ser[19], and if this reaction is only catalyzed by CaM–PKII (resulting in an increase in V_{max} with no change in the K_m for BH_4), we are left with a nagging question: What is the explanation for the reports that activation of TH by depolarization leads to a decrease in the K_m for the pterin cofactor (128, 129)?

The identification of the serine residues that are phosphorylated by various kinases (Table IX), when taken together with the knowledge of how activation of TH by these kinases is expressed (Table VIII), might lead to the conclusion that at physiological pH, phosphorylation of Ser[40] by any kinase would activate the enzyme by decreasing the K_m for the pterin cofactor and by increasing V_{max}, whereas phosphorylation of Ser[19] by CaM–PKII would activate by modestly increasing V_{max}.

There are indications, however, that such a conclusion would be wrong. Phosphorylation of pure TH from PC12 cells at Ser[40] by PKA, PKC, and CaM–PKII showed that the maximum phosphorylation of this site that can be achieved by these three kinases, surprisingly, is not the same and that under the assay conditions used (1 mM 6MPH$_4$, pH 6.1) only phosphorylation of this site by PKA activates the enzyme ($\sim$ threefold) (188). To put the lack of activation by PKC and CaM–PKII in perspective, it should be recalled that phosphorylation by PKC does activate the enzyme by decreasing the K_m for the pterin cofactor (assayed at pH 7) (158) (see Table VIII) and that phosphorylation by CaM–PKII does not activate in the absence of activator protein (143, 144).

The stoichiometry of phosphate incorporation at Ser[40] (mol/mol hydroxylase subunit) catalyzed by the three kinases was 0.78 by PKA, 0.43 by PKC, and 0.43 by CaM–PKII. In confirmation of earlier results (177), the last enzyme also catalyzed the incorporation of 0.76 mol at Ser[19] for a total phosphorylation of 1.19 mol/mol mediated by this kinase.

It was concluded from these results that PKA can phosphorylate all four of the subunits at Ser[40] (the results look more like three out of four subunits but the amount of endogenous phosphate in the enzyme, which was not determined, may account for this difference) and that PKC and CaM–PKII can only phosphorylate two of the four subunits.

When the hydroxylase was maximally phosphorylated by either PKC (0.49 mol/mol) or CaM–PKII (1.19 mol/mol) and then phosphorylated by PKA, an additional 0.30–0.36 mol/mol of phosphate was incorporated (leading to the same activation as seen with PKA phosphorylation alone), suggesting that almost one-half of the subunits were phosphorylated at Ser[40] by PKC or CaM–PKII. Moreover, when the enzyme was maximally phosphorylated by CaM–PKII and then incubated with PKC, no more phosphorylation was observed, an indication that these two kinases phosphorylate the same Ser[40].

To explain the observation that only phosphorylation of Ser[40] by PKA directly activates TH, it was suggested that the subunits of the hydroxylase may be in at least two different states. The subunits in one state are phosphorylated at this serine residue by all three kinases and this modification does not lead to activation; those in the other state are phosphorylated at this site only by PKA and this does lead to direct activation. Phosphorylation of Ser^{19} by CaM–PKII presumably is the one that requires activator protein to express the activation.

These results are somewhat reminiscent of those obtained with PAH, where hydrolysis of 75% of the total phosphate from the fully phosphorylated enzyme deactivates the enzyme to the same extent as hydrolysis of all of the protein-bound phosphate (189), an indication that one-quarter phosphorylated PAH (i.e., enzyme in which one subunit out of four is phosphorylated) is in the same state of activation as that of the unphosphorylated enzyme. With respect to Ser^{40} of TH, however, it appears, according to the data of Funakoshi et al. (188), as if the half-phosphorylated tetramer is in the same basal state (i.e., not activated) as is the non-phosphorylated enzyme.

Before examining the proposed two-state model, it is important to compare the data on which it is based with relevant published data. In a study of the relationship between the extent of phosphorylation of rat brain TH by PKA and the extent of activation (hydroxylase activity measured at pH 6.0 with 75 μM BH$_4$) (124), there was no indication that incorporation of the first 0.5-mol/mol of subunit is not associated with activation of the enzyme. Although the relationship did suggest a small allosteric effect, with the incorporation of the first 0.5 mol/mol activating less than that resulting from incorporation of the second 0.5 mol/mol, half-maximum phosphorylation did result in a four- to fivefold activation.

These results, therefore, do not support the assumption that phosphorylation of one-half of the subunits of TH by PKA does not activate the enzyme. It must be emphasized that Funakoshi et al. (188) did not test this assumption, that is, they did not determine whether the PKA-derived half-phosphorylated species is activated.

Can differences in the hydroxylase assay conditions used in the two studies account for the difference between the actual results (124) and the predicted ones (188)? In the former study, 75 μM BH$_4$ was used, whereas 1 mM 6MPH$_4$ was used in the latter one. The use of different pterins is unlikely to account for this discrepancy since activation by PKA has been observed with both 6MPH$_4$ and BH$_4$ (51, 52). It is possible, however, that the use of very different concentrations of the pterin cofactor in the two studies could be a crucial difference. Under the assay conditions used by Nelson and

Kaufman (124), in which 75 μM BH$_4$ was used, a level far below the K_m value of about 600 μM (see Table IV), activation of the enzyme that is expressed by a decrease in the K_m for the pterin would be much greater than that seen with 1 mM 6MPH$_4$, the conditions used by Funakoshi et al. (188) and could account for the different results in these two studies.

As discussed in the introduction, the first mechanism proposed to explain the acute activation of TH by increased neural activity assigned a central role to end product inhibition by catecholamines, such as norepinephrine and dopamine, and relief of this inhibition by a decrease in the intracellular level of these inhibitors. This notion was further elaborated by the demonstration that the inhibition by certain catechols is competitive with the pterin cofactor (100). Indeed, this characteristic provided an approach to test the validity of the simple end product regulatory mechanism, since it predicted that if activation of TH by nerve stimulation were due only to relief of inhibition by catecholamines, excess pterin should increase hydroxylase activity in the control tissue up to the level in the stimulated tissue. When this prediction was tested in the guinea pig vas deferens preparation it failed, an indication that the activation of the enzyme by nerve stimulation does not simply result from a decrease in end product inhibition (103).

The possibility that this kind of inhibition might nonetheless play a role in the regulation of the enzyme was bolstered by the finding that electrical stimulation of the vas deferens, as well as the addition of micromolar concentrations of Ca^{2+}, each activated the enzyme and increased the K_i of norepinephrine (106), and that electrical stimulation of dopaminergic neurons activated and increased the K_i for dopamine (129).

Activation of TH by cAMP in striatal extracts and synaptosomes also increased the K_i for end product catecholamines (113, 127, 129). These observations established a potential functional link between what might otherwise have appeared to be two unrelated regulatory processes: activation of the enzyme by relief from end product inhibition and activation by phosphorylation (i.e., activation by phosphorylation might result from decreased inhibition by catecholamines).

In support of this idea, Okuno and Fujisawa (190) reported that acid precipitation of crude striatal TH increased enzyme activity at neutral pH, an indication that an inhibitor was removed by this procedure. It was also shown that the incubation of the acid-treated enzyme with micromolar concentrations of catecholamines (e.g., dopamine or norepinephrine) inhibited the activity in a time-dependent manner and that phosphorylation by PKA of the dopamine-inhibited enzyme activated it (191). As pointed out by the authors, the relative slowness and lack of free reversibility of this dopamine-induced inhibition distinguishes it from the rapid, freely reversible competitive inhibition by catecholamines.

The link between phosphorylation and end product inhibition was further strengthened when it was demonstrated that pure bovine adrenal TH contains a blue-green chromophore, which is due to bound norepinephrine and epinephrine ($\sim$0.36 mol of catecholamine per mol of subunit) coordinated to the iron (Fe^{3+}) of the enzyme (192). This observation proved to be informative not only about the mechanism of action of the enzyme, but also about the enzyme's regulation. It has been shown that the Fe^{3+}-form of TH, which is presumably inactive, cannot be reduced to the active Fe^{2+} form by $6MPH_4$ when it is complexed with catecholamines. Phosphorylation of the enzyme by PKA (at Ser^{40}) facilitates reduction of the Fe^{3+} center under turnover conditions, that is, in the presence of both $6MPH_4$ and tyrosine, with the expected increase in enzyme activity (193). Phosphorylation also leads to a sixfold increase in the rate constant for dissociation of bound norepinephrine (194). These authors suggested that this phosphorylation-mediated increase in the rate of dissociation of the end product inhibitor can partly explain the activation of the enzyme by phosphorylation.

Studies of pure cloned TH that has been expressed in *Escherichia coli* have further clarified the way that bound catecholamines and phosphorylation interact to regulate the activity of the enzyme.

Since bacteria lack PKA and CaM–PKII (195), it was anticipated that the cloned enzyme isolated from *Escherichia coli* would not be phosphorylated and would consequently have very low activity. In fact, some investigators concluded that the nonphosphorylated form of the enzyme may have negligible activity under physiological conditions (58, 196).

It was therefore surprising that the cloned enzyme was not only highly active, but that the additional activation seen after phosphorylation catalyzed by PKA was modest (197–199). At pH 7.0–7.2, for example, phosphorylation decreased the K_m for BH_4 by 50–60% with no change in V_{max} (197–199), rather than the approximately 10-fold increase in V_{max} seen after phosphorylation of the striatal enzyme. Furthermore, when assayed at pH 6, phosphorylation only decreased the K_m for BH_4 by 30% with no change in V_{max}, rather than the expected sevenfold decrease in K_m (51) (Fig. 2). Another property that the cloned hydroxylase from *E. coli* shared with the enzyme activated by phosphorylation was its pH optimum, which was around pH 7.5 (197, 199), rather than around pH 6, the value that is characteristic of the unactivated enzyme (52, 57, 123). Also pointing to the cloned enzyme being activated, phosphatidyl-L-serine, which activates the native enzyme by decreasing the K_m for BH_4 severalfold (66), had essentially no affect on the K_m for the pterin at pH 6.3 (200). The attenuated response of the cloned enzyme to phosphorylating conditions was not due to a failure of the enzyme to be phosphorylated; on incubation with ATP and PKA, as much as 1 mol of phosphate was incorporated per mol of TH subunit (200).

The evidence that TH isolated from bovine adrenals and PC12 cells contains significant amounts of bound dopamine (192, 193), together with the likelihood that *E. coli* does not synthesize catecholamines such as dopamine, suggested that the cloned enzyme isolated from *E. coli* might be in an activated state because it contained no bound catecholamines. Incubation with stoichiometric amounts of dopamine converted the activated hydroxylase to a deactivated form whose catalytic properties are essentially the same as those of the native enzyme: at pH 7, a doubling of the K_m for BH_4 and a 50% decrease in V_{max}, as well as a shift in the pH optimum from about 6.7–5.7 (200). Finally, phosphorylation of the hydroxylase–dopamine complex resulted in activation that was almost indistinguishable from that seen after phosphorylation of the native enzyme, that is, the deactivating effect of dopamine was largely reversed (198, 200).

The dramatic changes in the catalytic properties of TH on incubation with dopamine result from binding a surprisingly small amount of the catecholamine to the enzyme. Although dopamine binds with high affinity ($K_d = 1 \ \mu M$), the stoichiometry is low (0.08 mol/mol of hydroxylase subunit, $M_r = 60,000$). The binding leads to the expected green color (201) ($\lambda_{max} = 660$ nm). In view of the likelihood that the catecholamine binds to the iron center of the enzyme, one of the reasons for the low stoichiometry observed in this study was probably the low iron content of the cloned enzyme (~ 0.4 mol/mol of subunit). In accord with this possibility, preincubation of the enzyme with exogenous Fe^{2+}, increased the stoichiometry of dopamine binding by 50% (200).

Despite the impressive recent advances in our understanding of the physiological regulation of TH by phosphorylation, reviewed above, it is still not possible to elaborate a comprehensive model describing how the enzyme is activated by various mediators. For a discussion of some earlier models see (128, 142, 202).

For a completely satisfactory model, it would be desirable to be able to describe a step-by-step sequence leading from the initial stimulus to the final result, namely, activation of the enzyme. Such a sequence can only be written for those activations that are mediated by agents such as forskolin, which increase the intracellular concentration of cAMP and thereby activate PKA. A cornerstone of this kind of activation is the probable functional link between phosphorylation of TH by PKA and decreased sensitivity to end product inhibition caused, in part, by an increase in the rate of dissociation of bound catecholamines from the phosphorylated enzyme. A likely sequence in this situation would be the increase in the concentration of cAMP, activation of PKA, enhanced phosphorylation of TH at Ser^{40}, decreased end product inhibition, and activation of TH. With respect to the last two steps, it should be recalled that evidence indicates that it is not the phosphorylation step per se that activates the enzyme, but rather the resultant decrease in end product inhibition (198, 200).

For one of the physiologically most important modes of activating TH, namely, activation induced by electrical or chemical depolarization of neural tissue, essential details about what is going on are lacking. Consider the following reasonable sequence: depolarization, influx of Ca^{2+}, efflux of catecholamines (162), decreased end product inhibition, enhanced phosphorylation of TH [probably at Ser^{19} and perhaps also at Ser^{31} (see Table IX)], and activation of TH (in the presence of activator protein). Problems abound. The most glaring problem is that it is not certain which step in the sequence actually leads to activation of the enzyme. Is it the phosphorylation step or the efflux of catecholamines? Moreover, as already discussed (see Table IX), although CaM–PKII is probably the enzyme responsible for the *in situ* phosphorylation of Ser[19], the kinase that phosphorylates the other site (Ser[31]) in response to depolarization has not been identified. It may be recalled that ERK1 and ERK2 can phosphorylate this site but there is no evidence that these kinases are themselves activated by depolarization. A further complication is the finding that depolarization decreases the sensitivity of TH to dopamine-induced inhibition (129, 203). This is a complication because unlike phosphorylation of TH by either PKA (113, 127, 129) or PKC (158), which results in an increase in the K_i for dopamine, phosphorylation of the hydroxylase by ERK1 has been reported to have no effect on the K_i for dopamine (184). As for phosphorylation by CaM–PKII, the results are not straightforward: Phosphorylation by this kinase of one of the two predominant isoforms of human brain TH (TH-1) does not affect the K_i value for dopamine, whereas there is an increased with phosphorylation of isoform TH-2 (197). These results indicate that phosphorylation of the hydroxylase at Ser[19] and Ser[31], most likely catalyzed by CaM–PKII and ERK1, respectively, may not be able to fully account for the increase in the K_i for dopamine seen after *in vivo* activation of TH following electrical depolarization. These considerations raise the possibility that phosphorylation of TH by either PKA or PKC may yet be found to play secondary or indirect roles in the depolarization-induced *in situ* activation of the hydroxylase.

With respect to the possibility that PKA-mediated phosphorylation and activation of TH may be involved in this process, Fujisawa et al. (204) proposed that in addition to the well-established role of CaM–PKII (and activator protein), PKA is also involved in the depolarization-mediated activation of the enzyme. This proposal is based on the assumption that during depolarization of the nerve terminal, the ATP that is released together with catecholamines is degraded to adenosine, which promotes the accumulation of cAMP in the cell via adenosine receptors, thereby activating PKA, which, in turn, would mediate the phosphorylation activation of TH. The postulate is supported by the observation that adenosine is a potent activator of the hydroxylase in PC12 cells (205, 206).

The proposal by Fujisawa et al. (204) has attractive features but it cannot be readily reconciled with some facts. In particular, the catalytic properties of activated TH isolated from tissues after membrane depolarization are incompatible with those expected for TH activated *in vitro* by PKA-catalyzed phosphorylation. Thus, as already discussed, although a few early studies reported that depolarization-mediated activation of the enzyme was associated with a decrease in the K_m for the pterin cofactor with no change in V_{max} (106, 128, 129), changes that are compatible with those mediated by PKA, other studies have not observed a change in the K_m for the cofactor (103, 138, 203, 207). Also incompatible with a direct phosphorylation of TH by PKA is the failure to observe an alkaline shift in the optimum pH following depolarization (138, 203, 208), a change that is characteristic of TH after phosphorylation that is mediated by PKA (57). Finally, although prolonged (20 min) *in vivo* electrical stimulation of afferent dopaminergic fibers increases the phosphorylation of Ser[19], Ser[31], and Ser[40] (182), KCl- or veratridine-induced depolarization of striatal synaptosomes (182), or PC12 cells (183) increases the phosphorylation of Ser[19] and Ser[31] but not of Ser[40]. Both the failure to observe any *in vitro* depolarization-induced phosphorylation of Ser[40], the only serine residue that is phosphorylated by PKA, together with the failure of depolarization to produce the kinetic changes in TH that are characteristic of the PKA-activated enzyme, make it unlikely that this kinase plays a direct role in the depolarization-induced activation of the enzyme.

In addition to the gaps in our knowledge that have just been outlined, there are further complexities that must be accommodated by any model for the physiological regulation of TH. There are indications, for example, that the *in situ* activation of the enzyme may occur in stages, with some changes in its catalytic properties being discernable before others. Weiner et al. (209) found that in the stimulated hypogastric nerve–vas deferens preparation, an increase in V_{max} of the hydroxylase preceded a change in K_m for the pterin cofactor. Another indication of a multistep process is the observation that following depolarization of PC12 cells by 56 mM KCl, phosphorylation of TH (maximum in ~1 min) precedes its activation (maximum in ~3 min) (207). Also pointing to at least a two-stage process involved in the activation of TH, Iuvone et al. (210) reported that in light-stimulated retinal amacrine neurons, a decrease in the K_m for the pterin cofactor occurs much sooner than an increase in V_{max}.

Finally, certain drugs, such as gamma-hydroxybutyrate, appear to activate TH *in vivo,* but the activated state does not persist *in vitro* [reviewed in (85)]. These effects may represent an extreme example of a multistep activation with only an early, reversible step being stimulated by the drug.

In an attempt to explain findings such as these, a model for the *in situ* regulation of TH was proposed (85) that assumes the initial event leading to the relatively persistent activation of the enzyme in a stimulated tissue, such as a dopaminergic neuron, is a change in the enzyme's environment, rather than in the covalent structure of the enzyme itself. This environmental change thus leads to a freely reversible *in situ* activation. The nature of this change is not known with certainty, but a likely candidate, and one that was proposed quite early (96, 101, 211), would be a decrease in inhibition of the enzyme by dopamine that would result from the depolarization-induced release of dopamine. Since Ca^{2+} is involved in this type of activation of the hydroxylase, as well as in the release of dopamine, the triggering event could be, as others have suggested (106), a depolarization-stimulated entry of Ca^{2+} into the neuron. If the proposed environmental change is the release of dopamine, the resulting activation of TH would be expected to persist *in vitro*. Other possible activating *in situ* changes might not be. In certain tissues, in which the necessary protein kinases are present, this stage would be followed by the second stage involving phosphorylation of the hydroxylase. It is only the occurrence of this second stage in which the enzyme is covalently modified that the activated state would be invariably fixed so that the higher activity could be detected *in vitro*. Although this model assumes that the enzyme is activated in each of the two stages, the kinetic expression of the activated state is not necessarily the same for each one.

The attractiveness of such a model would be enhanced if there was a causal connection between the two postulated stages of activation. In 1985, when this proposal was made, there was no indication of how these two stages might be linked. To connect them, it was postulated that "activation of the enzyme leads to, or facilitates, its phosphorylation (85)." That is, phosphorylation of TH not only decreases end product inhibition, but, in addition, decreased inhibition reciprocally increases phosphorylation.

This postulate has recently received some experimental support. Daubner et al. (198) demonstrated that the rate (but not the extent) of phosphorylation of a rat TH–dopamine complex by PKA is about 50% the rate with the free hydroxylase. Similar, if less marked, effects of added dopamine were reported for the rate of the PKA-catalyzed phosphorylation of recombinant human TH; in this study, dopamine also modestly decreased the extent of phosphorylation (212). Whether this connection between dopamine and the rate of phosphorylation of TH can be extended to the depolarization-mediated activation of the enzyme would depend on whether or not dopamine can also inhibit phosphorylation of the hydroxylase by CaM–PKII.

The two-stage model for TH activation bears some resemblance to current views on the acute regulation of hepatic PAH. With the latter enzyme, BH_4 and phenylalanine affect the rate of PKA-mediated activation–phosphorylation in opposite ways, with BH_4 inhibiting and phenylalanine stimulating (213, 214). In an extension of this kind of study to pure rat brain TH, neither BH_4 nor tyrosine was found to have any effect on the rate of phosphorylation of the enzyme catalyzed by PKA. On the other hand, although tyrosine had no effect, BH_4 and other tetrahydropterins, in concentrations as low as 5 μM, markedly accelerated the dephosphorylation of TH by the major phosphatase detected in rat caudate and partially purified from rat tissue (215). Since the phosphorylated TH that was used in these studies had been phosphorylated by PKA, the residue that had been phosphorylated and that the phosphatase had dephosphorylated was presumably Ser^{40}. The stimulation by BH_4 is due to an approximately six- to eightfold increase in V_{max}, with little, if any change in the K_m of the phosphatase for TH, which remained at 40–60 pM. The exact nature of the BH_4-responsive phosphatase is not known but the fact that the phosphatase assays were carried out in the absence of Ca^{2+} and in the presence of EGTA indicates that it is not a 2B type that is stimulated by Ca^{2+} and calmodulin (215). Rather, data to be discussed below support the conclusion that it is a 2A type phosphatase.

In addition to the stimulation of this phosphatase by pterins, micromolar concentrations of GTP, the biological precursor of BH_4, strongly inhibited the enzyme, 50% inhibition being observed at less than 1μM (215). Adenosine triphosphate also inhibited but was less potent than GTP. The physiological significance of this inhibition is not known.

The possibility that the BH_4 stimulation of the phosphatase occurs *in situ* was studied in rat striatal synaptosomes (216). To enhance the likelihood that dephosphorylation of the same serine residue (i.e., Ser^{40}) was being studied in synaptosomes, TH was labeled with $^{32}P_i$ in the presence of a cAMP analog. Treatment of the ^{32}P labeled synaptosomes with BH_4 decreased the incorporation of ^{32}P into TH, with a 50% decrease occurring at 230–250 μM and the half-maximum response seen at about 100 μM. Okadaic acid, a rather specific inhibitor of type 2A phosphatase, almost completely prevents the stimulation by BH_4, an indication that it is a type 2A phosphatase that BH_4 stimulates (76). This finding is coherent with results of Haavik et al. (217) who concluded that type 2A phosphatase accounts for approximately 90% of the TH phosphatase activity in both adrenal medulla and in striatal tissue.

The physiological significance of this effect of BH_4 on the state of phosphorylation, and, therefore, on the state of activation of TH, is not known. Since it is likely that the hydroxylase functions in the brain at tissue concentrations of BH_4 that are well below saturating, so that increased levels of BH_4 would increase hydroxylase activity, the stimulation by BH_4 of the dephosphorylation of TH would tend to insulate to some extent the activity of the enzyme—and therefore of catecholamine synthesis—from changes in tissue levels of BH_4.

It should be noted that with both hepatic PAH and striatal TH, BH_4 decreases the state of phosphorylation and the extent of activation of the enzymes, but the mechanism by which it accomplishes this is different: with PAH, BH_4 inhibits phosphorylation of the enzyme, whereas with TH it stimulates the dephosphorylation.

As already mentioned, the two stage model for TH activation is somewhat analogous to the way in which PAH is believed to be acutely regulated. The similarities can be seen in the following parallel schemes for the two enzymes, where PAH* and TH* represent transiently activated species, and PAH-P and TH-P, represent phosphorylated species of the two enzymes; DA is dopamine and Phe is phenylalanine.

$$PAH \underset{BH_4}{\overset{Phe}{\rightleftharpoons}} PAH^* \overset{Phe}{\underset{BH_4}{\nrightarrow}} PAH^*\text{-}P$$

$$TH \underset{DA}{\overset{BH_4}{\rightleftharpoons}} TH^* \underset{BH_4}{\overset{?}{\rightleftharpoons}} TH^*\text{-}P$$

With each enzyme, there is an activated species that does not involve a covalent modification (designated PAH* and TH*) and an activated species that does, that is, the phosphorylated species. With both enzymes, the first postulated stage of activation is regulated by the opposing effects of ligands: With PAH, the activation by phenylalanine overcomes the deactivating effect of BH_4, whereas with TH, BH_4 can overcome the inhibition by dopamine; that is, with TH, removal of dopamine is the functional equivalent of activation of PAH by phenylalanine. As can be seen, although BH_4 is involved in the noncovalent activation of both enzymes, with PAH it plays a negative role, whereas with TH it plays a positive role. Furthermore, with both enzymes formation of the noncovalently activated species appears to facilitate their conversion to the covalently activated forms via phosphorylation. What is not depicted in the scheme is the fact that the two stages of activation also interact reciprocally: With PAH, phenylalanine not only facilitates phosphorylation (213, 214), but phosphorylation facilitates activation by phenylalanine (214, 218). With TH, removal of dopamine not only facilitates phosphorylation (198), but phosphorylation accelerates the dissociation of inhibitory catecholamines (194).

VII. Induction of Tyrosine Hydroxylase

As discussed previously, TH can be acutely activated (within the time frame of seconds to minutes) in response to a variety of signals by enzymatic phosphorylation of existing hydroxylase molecules. In addition to this short-term regulation, the hydroxylase can also be modulated by long-term adaptation (occurring in hours to days) to various environmental stimuli. The latter mode of regulation involves increased synthesis of new hydroxylase molecules, that is, enzyme induction.

The common denominator of many of the stimuli that can elicit this long-term response *in vivo* [in adrenals and/or in central or peripheral adrenergic neurones such as superior cervical ganglia (SCG)] is that they all increase the utilization of intracellular catecholamines. This depletion of neurotransmitters not only leads to an immediate compensatory short-term activation of TH, which serves to replenish depleted stores of the transmitter, but also leads to a long-term adaptative induction of the hydroxylase.

Among the effective stimuli are many of the same ones that are also involved in the short-term activation of TH and, as will be discussed, these long-term effects also appear to involve the same second messenger molecules, such as cAMP and Ca^{2+}. *In vivo,* these stimuli include cold stress (219), immobilization stress (220), prolonged nerve stimulation (221), and certain catecholamine-depleting drugs, such as reserpine, that can pharmacologically mimic the effects of stress (222). Induction of TH has also been demonstrated in cultures of adrenal chromaffin cells, pheochromocytoma cells, and SCG by a variety of effectors including glucocorticoids (223, 224), NGF (225–227), EGF (228) prolonged nerve stimulation (229, 230), and cAMP (225, 231).

This kind of delayed increase ($\sim$ two to threefold) in TH activity was first observed with the adrenal enzyme in rats that had been subjected to chemical sympathectomy by treatment with 6-hydroxydopamine (232), an agent that selectively destroys adrenergic nerve endings (233). A similar effect on adrenal TH was reported in rats 24–48 h after treatment with other drugs that interfere with postganglionic sympathetic transmission. Reserpine, which, as mentioned earlier, depletes cellular stores of catecholamines, was found to increase TH activity in adrenals, SCG, and brainstem. Phenoxybenzamine, which blocks adrenergic receptors of the effector organs, also led to a delayed increase in adrenal TH activity (222, 234). Since these drug-mediated effects on the adrenal enzyme were prevented by a section of the splanchnic nerves supplying the adrenal glands, this delayed compensatory increase in TH activity was called transsynaptic induction of the enzyme (234).

At the time this name was coined there was better evidence for the "trans-synaptic" part of it than for the "induction" part. The finding that the increased hydroxylase activity was not due to a change in the K_m for either tyrosine or the pterin cofactor (222) was consistent with, but did not prove, that the increase was due to an increased rate of synthesis of the enzyme. Final proof that transsynaptic induction resulted in the increased synthesis of TH came from the immunochemical studies of Joh et al. (235), who demonstrated increased accumulation of hydroxylase protein in the sympathetic ganglia and the adrenal medulla following administration of reserpine to rats, and from Hoeldtke et al. (236), who showed that exposure of rats to cold, as well as immobilization stress, lead to increased amounts of adrenal TH protein as determined by immuno-titration with a specific antibody to the enzyme. Chuang and Costa (237) also reported that the incorporation of (^{3}H)leucine into protein that could be precipitated by a specific antibody to TH is greater in cold-stressed animals.

Early work on trans-synaptic induction focused on the peripheral ganglia and the chromaffin cells of the adrenal medulla. Reis et al. (105) examined the effect of reserpine in the central nervous system to determine whether the reported increase in TH activity in the *locus coeruleus* after reserpine or cold stress (238), was due to accu-mulation of the enzyme protein or activation of existing enzyme. Immunoprecipitation studies showed a two- to threefold increase in TH protein in the *locus coeruleus* 4 days after reserpine treatment. This increase was localized to the cell bodies of the noradrenergic neurons. In contrast, a smaller increase in TH was observed in the hypothalamus, a region rich in noradrenergic nerve terminals. Since trans-synaptic induction is due to the synthesis of new hydroxylase molecules and since protein synthesis occurs in the cell body, it is not surprising that this is where the greatest stimulation was ob-served. The smaller stimulation in the noradrenergic nerve terminals, areas that have 40–100 times more TH than the cell bodies, probably represents the transport of a relatively small amount of new enzyme into a large preexisting pool. Dopaminergic structures were also ex-amined and no change in TH activity could be observed in the *sub-stantia nigra* or in the caudate nucleus, structures that contain the cell bodies and the nerve terminals, respectively, of the dopami-nergic nigrostriatal pathway. The action of reserpine on TH in the central nervous system appears to be localized primarily in the cell bodies of noradrenergic neurons with small increases in noradrener-gic nerve terminals.

Thoenen and Otten (239) demonstrated that the common feature of all of the experimental conditions that produce trans-synaptic in-duction of TH is an increased activity of the preganglionic choliner-gic nerve leading to the liberation of acetylcholine. Since the action of acetylcholine on the nicotinic receptor of the adrenergic neuron is believed to be the event that initiates trans-synaptic induction, they proposed that acetylcholine is the first messenger in the series of events leading to induction of TH.

Guidotti et al. (240) and Kurosawa et al. (165) provided evidence that both acetylcholine and carbamylcholine lead to an increase in cAMP in the adrenal medulla. Costa et al. (241) proposed that this increase in cAMP initiates a sequence of biochemical events that culminates in the induction of TH in the adrenal medulla. The steps in the proposed sequence are (1) the activation of a cytoplasmic PKA, (2) the translocation of the catalytic subunits of PKA from the cytosol to the nuclear fraction, (3) increased phosphorylation of protein tightly bound to chromatin, (4) increased mRNA synthesis, and (5) increased synthesis of TH. Evidence has been presented to substantiate the involvement of each of these steps. Continued nerve stimulation is required (242) for this sequence of events to culminate in the induction of TH.

Thoenen and Otten (239) questioned the role of cAMP in the trans-synaptic induction of TH in the adrenal medulla. They cited the evidence that in the denervated adrenal medulla, aminophyline produced a large increase in cAMP but no increase in TH activity. Costa et al. (243) replied that, under these conditions, there is no translocation of the protein kinase from the cytosol to the nuclear fraction. Moreover, Costa et al. (243) emphasized that it is not the increase in cAMP but the subsequent activation and translocation of the protein kinase that is the important biochemical signal leading to the induction of TH in the adrenal medulla. In support of this hypothesis, it should be noted that similar activation and translocation of protein kinase has been described in other tissues [see (244, 245)].

In the adrenal medulla, the evidence supports a role for the activation and translocation of protein kinase in the induction of TH. However, the necessity for continued cholinergic innervation for induction to take place does not, as yet, have a clear biochemical explanation. A better understanding of this aspect of the induction of TH may help to resolve some questions raised by Thoenen and Otten (239) concerning the role of cAMP in this complicated process.

The induction of TH in the SCG, like that in the adrenal medulla, requires nerve stimulation and the participation of nicotinic receptors; however, in many other respects it differs from the process in the adrenal medulla (246). In the SCG, but not in adrenal medulla, muscarinic receptor agonists block the induction of TH by nicotinic receptor agonists. Moreover, atropine, a muscarinic antagonist, facilitates the ability of nicotine to induce TH not only in intact, but also in decentralized SCG. Part of the difference between the SCG cervical ganglion and the adrenal medulla can be explained by the fact that in the SCG cervical ganglion, stimulation of muscarinic receptors activates the release of dopamine from the small intensely fluorescent (SIF) cells, whereas in the adrenal medulla there are no interneurons that regulate the chromaffin cells. That the release of dopamine from the SIF cells inhibits the induction of TH has been confirmed by the finding that in the SCG ganglion, apomorphine, a dopamine agonist, also antagonized the induction of TH by nicotine (246).

Glucocorticosteroids such as dexanethasome can induce TH in SCG, but not in the adrenal medulla (246). This induction requires intact innervation and can be blocked by cortexolone, a compound without hormone activity that specifically prevents the binding of glucocorticoids to the cytosolic steroid receptor.

In contrast to the adrenal medulla, TH can be induced in the SCG by β-receptor agonists such as isoproterenol and epinephrine. This induction can take place in decentralized as well as intact SCG. Finally, there seems to be general agreement, again in contrast to the adrenal medulla, that an increase in cAMP does not usually precede the induction of TH in the SCG.

As mentioned previously, the induction of TH has also been studied in tissue culture. Waymire et al. (247) reported that the addition of dibutyryl-cAMP, papaverine (a phosphodiesterase inhibitor), or sodium butyrate to mouse neuroblastoma cells in culture could increase the activity of the hydroxylase 15–50-fold. In addition, glucocorticoids, such as dexamethasone, were also effective (248). Testosterone and estradiol-17β were without effect. The effect of simultaneous treatment with glucocorticoids and cAMP analogs was found to be greater than the sum of the increases produced by either agent alone (249). That these increases in activity were due to increases in the number of TH molecules was shown by immunotitration. The workers also confirmed that progesterone, testosterone, and β-estradiol did not induce TH. However, progesterone, an antagonist of the cytosolic glucocorticoid receptor, inhibited the dexamethasone induction of the hydroxylase. Although no increased incorporation of (^{3}H)uridine into total RNA in the presence of dexamethasone or dexamethasone plus a phosphodiesterase inhibitor was observed, the induction of TH could be inhibited by the simultaneous addition of minimal amounts of actinomycin D to the culture.

These results suggested that the glucocorticoid-mediated induction of the hydroxylase required RNA synthesis. That there is a selective increase in mRNA can be inferred from the work of Baetge et al. (250). These workers purified polysomal polyA-mRNA from a rat pheochromocytoma cell line before and after treatment of the cells with dexamethasone. This purified mRNA was then translated in a cell-free protein synthesizing system and the TH identified by immunotitration. A 2.5-fold increase in the ratio of TH to total protein in this system was observed after treatment of the cells with dexamethasone. This finding corresponded exactly to the increase of TH to total protein observed in the culture as measured by the incorporation of (^{3}H)leucine into protein followed by immunoprecipitation of TH. Double-label experiments were used to rule out an effect on protein degradation. These studies demonstrated that the increased synthesis of TH following dexamethasone is the result of an increased relative amount of TH mRNA.

Evidence from different sources is consistent with the idea that the action of steroid hormones is mediated through binding of the hormone to a cytosolic receptor. The hormone–receptor complex has an increased affinity for chromosomal sites in the nucleus where it triggers the characteristic response (251). The results of Tank and Weiner (249) suggest an interaction between the glucocorticoid hormone and cAMP. The interaction is not direct since dexamethasone does not appear to increase cAMP levels. Rosenfeld and Barrieux (252) reported that both dexamethasone and cAMP increase the amount of functional mRNA for several proteins and proposed that this action is both synergistic and sequential. In the case of tyrosine aminotransferase, for example, there is evidence that the action of glucocorticoids on mRNA transcription precedes the effect of cAMP. Wicks et al. (253) also reported that steroid hormones control the level of the cytoplasmic cAMP-dependent protein kinase type I. Adrenalectomy or hypophysectomy, for example, will significantly decrease the amount of cAMP-dependent protein kinase type I, the kinase that Costa et al. (243) proposed is translocated from the cytosol to the nucleus, would offer a second means of interaction between steroid hormones and cAMP. Tank and Weiner (249) suggested that the interaction between glucocorticoid hormones and cAMP seen in the neuroblastoma cell line may be analogous to the glucocorticoid enhancement of trans-synaptic induction of TH.

Thoenen et al. (254) reported that treatment of newborn rats with NGF resulted not only in the expected morphological changes but also in a selective induction of TH and dopamine β-hydroxylase in SCG. The specific activity of TH in the SCG of NGF-treated animals was 5.4-fold higher than in control animals. Only slight increases in the specific activities of dopa decarboxylase and monoamine oxidase were seen after NGF treatment. Since mixing experiments (control plus treated) provided evidence against the formation of an activator, they proposed that this increase in specific activity represented an increase in the amount of enzyme protein.

Thoenen et al. (254) pointed out the similarities between induction of TH by NGF and the trans-synaptic induction of TH and dopamine β-hydroxylase by cold exposure or treatment with reserpine. Hendry and Iverson (255) also demonstrated that NGF produces an increase in TH in SCG. Moreover, they showed that injection of the antiserum to NGF produced a profound, dose-related inhibition of the normal increase in TH in the SCG over the first 10 days of life.

In rat sympathetic ganglia in organ culture, glucocorticoids were found not only to induce TH but also to modulate the induction of the enzyme by NGF (256). Previous work with the same preparation had demonstrated that decentralization prior to organ culture nearly abolished the ability of dexamethasone to induce TH (257). Decentralization, however, did not significantly affect the ability of NGF to induce the enzyme or the ability of corticosterone to potentiate this effect. Like the induction by dexamethasone (257), the induction by NGF or NGF plus corticosterone was abolished by cycloheximide.

In both the intact and the decentralized preparation, the effects of NGF plus corticosterone on TH activity were greater than the sum of the two separate effects. Corticosterone increased the maximum response as well as the response at low levels of NGF. Perhaps the most remarkable effect of corticosterone was on the time course of induction of TH by NGF. Without preincubation with corticosterone, the time for maximal induction of the hydroxylase was 4 h; with preincubation with corticosterone, the time for maximal induction by NGF was reduced to 10 min. The authors stated that this was true whether intact or previously decentralized ganglia were used. This dramatic shift in the time course further strengthens the argument that glucocorticoids act synergistically with NGF. That this effect was specific for glucocorticoids was demonstrated by the lack of effect of such steroids as estradiol, protesterone, and testosterone. Moreover, the effect of corticosterone on NGF induction of TH could be blocked both by progesterone and by cortexolone. Since progesterone is known to compete with corticosterone for glucocorticoid receptors (258) and cortexolone is a specific blocker of glucocorticoid receptors (259), the authors concluded that the interaction of glucocorticoids with the glucocorticoid receptors plays an important role in mediating and potentiating the induction of TH in the SCG.

VIII. Mechanism of Transcriptional Regulation

Although these early studies on the induction of TH generated a wealth of descriptive information about the kinds of signals that can lead to this long-term adaptation of the enzyme, the molecular mechanisms involved remained largely unexplored.

The isolation in the early 1980s of cDNA clones coding for TH (260, 261) provided a powerful tool for the study of the regulation of the synthesis of the hydroxylase. With the use of these clones, it was demonstrated that mRNA levels for TH increased following treatment of PC12 cells with derivatives of cAMP and glucocorticoids (261). Levels of TH mRNA were also elevated in adrenals, SCG, and in areas of the brain known to be rich in TH (e.g., *substantia nigra* and *locus coeruleus*) in rats after they had been treated with reserpine or subjected to cold stress (262–265).

Following these observations, numerous studies addressed the question of which regions of the *TH* gene regulate the induction of the enzyme. With the use of nuclear run-on assays, Lewis et al. (261) showed that the transcriptional activity of the *TH* gene is rapidly (within 10 min) increased in PC7e pheochromocytoma cells (derived from PC12 cells) after the cultures had been treated with dexamethasone or with agents such as forskolin or 8-bromo-cAMP, which increase intracellular levels of cAMP. It should be noted that this increase in the rate of TH transcription occurs long before an increase in TH mRNA levels are detectable ($\sim$3–4 h) (261, 266) or increases in TH activity are observable ($\sim$8–16 h) (261, 266). It is likely that this sequence of events reflects the respective turnover times of TH mRNA and TH proteins.

To analyze the regions of the *TH* gene that might mediate the response to these two inducers, the 5′ flanking sequences of the hydroxylase gene were fused to the gene for bacterial chloramphenicol acetyltransferase (CAT). Since transcription of the *CAT* gene in this hybrid is under the control of the regulator elements in the *TH* gene, the *CAT* gene functions as a reporter gene, reporting information about the sequences in the gene that contain promoter and/or enhancer elements. Studies carried out with pheochromocytoma cells and pituitary cells that had been transfected with the hybrid gene, showed that a region of the *TH* gene containing bases -272 to $+27$ (i.e., a region spanning 272 bases of the 5′ flanking sequence containing the transcription initiation site and 27 bases of transcribed nontranslated sequences) or bases -773 to $+27$ conferred induction of CAT by cAMP. By contrast, only a modest increase in CAT activity was observed when $+27/-273$ containing cultures were treated with dexamethasone, whereas the same treatment of cultures containing $+27/773$ actually depressed CAT activity below the basal level. These results showed that whereas the sequences required for the cAMP-mediated induction of TH are located within 272 bases of 5′ flanking sequence, those required for significant induction by glucocorticoids are not contained even within 773 bases of this region (267).

The location of a cAMP responsive element (it is not known whether it is the sole element) within the $+27/-272$ region is coherent with other results showing that a number of genes that are regulated by cAMP, such as the phosphoenolpyruvate carboxylase gene (268), contain a homologous core sequence (TGACGTCA) that is also present at -44 to -37 in the *TH* gene.

The failure to observe a meaningful glucocorticoid-mediated increase in CAT activity was unexpected since sequences -454 to -443 of the hydroxylase gene are homologous to the sequences that comprise the glucocorticoid regulatory element of the human metallothionein gene (269). As suggested by Lewis et al. (267), these negative results indicate that despite the sequence homology with the glucocorticoid regulatory element of the metallothionein gene, the sequence(s) in the *TH* gene responsible for the glucocorticoid induction of the enzyme has not yet been located.

Further analysis of the 5' flanking region of the *TH* gene in pheochromocytoma cells showed that progressive deletion of sequence -272 to -212 upstream from the transcription initiation site did not decrease transcriptional activity (as measured by the *CAT* reporter gene) but that deletion of additional sequences between -212 and -187 resulted in a drastic decrease in transcription (270). This essential region contains an AP-1 binding site (TGATTCA) between -206 and -200. The AP-1 (activator protein -1) has DNA-binding properties that are similar to the products of the *jun* and the *fos* oncogenes. In fact, these products may be the principal components of the AP-1 complex. The AP-1 binding sites occur in the control regions of many viral and cellular genes that are stimulated by treatment of cells with phorbol esters, thereby implicating PKC in the regulation of these gene. For this reason, the AP-1 binding sequence has also been called "TRE" for tetradecanoly phorbol acetate (TPA)-responsive element [for a review, see (271)].

With respect to the finding of potential an AP-1 binding site in the *TH* gene (270), and its relationship to the effects of phorbol esters in gene regulation, Mallet and co-workers (272) demonstrated that treatment of PC12 cells with TPA increases *TH* gene transcription. Treatment of some cells with TPA also increased the level of AP-1 (273). Further strengthening the link between *c-fos* and the induction of TH, reserpine, known to increase transcription of TH (262), also increases *c-fos* expression in adrenals and SCG of treated rats, with the expression of the hydroxylase being directly correlated with expression of *c-fos* (274).

The AP-1 site and *c-fos* have also been implicated in the NGF-mediated induction of TH in PC12 cells. Gizang-Ginsberg and Ziff (275) showed that *c-Fos* is one of the early genes induced by NGF and that the product of this gene can bind to the TH AP-1 site. The resulting stimulation of transcription of the hydroxylase gene increases the synthesis of the hydroxylase, which is crucial for neuronal differentiation.

In addition to the AP-1 binding site, the 5' flanking region of the *TH* gene contains several other potential-binding sites for other transcription factors, including an AP-2 site at -221, a POU/Oct site at

− 176, an SP-1 consensus sequence at − 120, and a cAMP response element (CRE) at − 45, as well as the TATA box sequence at − 24 to − 29 (270, 276). Since, as mentioned earlier, deletions up to − 187 markedly decreased transcription despite the presence of a CRE at − 45, it has been concluded that the CRE alone contributes little to the basal transcriptional activity of TH (270), although it does appear to mediate a large increase in transcription in response to forskolin (277). Surprisingly, it has been found that a 70 base pair (bp) region (− 229 to − 160], which contains potential-binding sites for all of the transcription factors mentioned above, as well as for E2A/MyoD (E box), when fused to CAT, confers forskolin inducibility in a pheo-chromocytoma cell line despite lack of an authentic CRE in the 70 bp fragment (277). Gel-shift assays showed that this fragment forms a cell type-specific complex with nuclear extracts from cells that express TH. Competition experiments with oligonucleotides containing binding sites for various transcription factors indicated that AP-1, AP-2, and E box sites, but not the POU site, may be involved in forming the cell-specific complex (277). Gel-shift assays with nuclear extracts from untreated and phorbol ester-treated cells also suggested that a protein complex binds to the AP-1 sequence (274).

The response of TH to nerve stimulation follows the same pattern as the response to most other stimuli: short-term stimulation leads to acute activation of existing enzyme molecules, whereas protracted nerve stimulation or chemical membrane depolarization leads to increased synthesis of new enzyme molecules.

Prolonged membrane depolarization elevates TH mRNA levels *in vivo* and in cell culture (230, 278–281). In PC12 cells, the depolarization-mediated increase in TH mRNA levels was shown to be inhibited by removal of Ca^{2+} with EGTA (282), an indication that a Ca^{2+}-dependent protein kinase might be involved.

The possibility that the responsible kinase might be PKC was made unlikely by the finding that neither sphingosine nor leupeptin, two antagonists of this kinase, inhibited the increased transcription induced by depolarization (283).

In PC12 cells transiently transfected with a plasmid containing the 5′ flanking sequence of the *TH* gene fused to the gene for CAT (− 773/ + 27), treatment with the depolarizing agent veratridine for 12 h increased CAT induction approximately threefold. The depolarization-responsive element was further delineated by deletion of additional sequences from the 5' flanking region. No decrease in induction of CAT was noted when the AP-1 site at position -200 to -206 and a sequence at -205 to -193 known to be important for the NGF-mediated regulation of transcription of the *TH* gene (275) were deleted (283). Moreover, a plasmid missing putative POU/Oct and SP-1 site was even more responsive to membrane depolarization than plasmid $-773/+27$. The shortest sequence that was still be able to respond to depolarization was a plasmid with 60 nucleotides upstream containing the CRE. It was not until part of the CRE was deleted (-45 to -38) that veratridine-mediated induction was lost, making it likely that this region contains the depolarization-responsive element. That the $-60/+27$ plasmid with its intact CRE retained the ability to respond to Ca^{2+} was supported by the observation that treatment of PC12 cells containing this plasmid for 12 h with the calcium ionophore ionomycin led to a fivefold increase in CAT activity (283).

The finding that the depolarization-responsive element mapped to the region of the *TH* gene containing the CRE raised the possibility that the depolarization effect might be mediated by cAMP. A small synergistic effect on CAT activity was observed when cells transfected with the 5' TH CAT construct ($-60/+27$) were treated with a combination of veratridine (or elevated KCl) and 8-bromo-cAMP (283). The mechanism of this kind of interaction, however, was not clarified by these results.

Implicit in the findings that agents that increase intracellular levels of cAMP (such as forskolin and certain cAMP derivatives) induce TH and that the induction of the enzyme by membrane depolarization also appears to involve the CRE, is the idea that PKA may play a role in mediating these transcriptional responses. Along the same lines, the evidence that Ca^{2+} is essential for the depolarization-stimulated induction of TH suggests that a Ca^{2+}-dependent kinase, such as CaM–PKII, may also be involved. That PKA does indeed play an important role both in the basal and the cAMP-inducible expression of the hydroxylase gene in PC12 cells was demonstrated by Kim et al. (284), who showed that in several PKA-deficient cell lines derived from rat PC12 cells, the basal expression of TH was significantly decreased (to 40–45% of the level in wild-type cells). This decrease was indicated by lower levels of the enzyme measured by hydroxylase activity and immunoreactivity, as well as by decreased steady-state levels of TH mRNA. In addition to the defect in basal expression, TH could no longer be induced by dibutyryl-cAMP in the deficient cells.

The effect on transcriptional activity of cotransfection of the PKA-deficient cells with an expression plasmid for the catalytic subunit of PKA (PKA_c) together with a plasmid containing the 5' flanking region of the *TH* gene (bp -2400 to $+27$) fused to the *CAT* reporter gene was studied. The PKA_c expression vector led to a 13–19-fold increase in transcriptional induction in both control PC12 cells and in the PKA-deficient cells. Although the authors stated that transfection with PKA_c "fully reversed" the transcriptional defect, the results do not entirely support this contention. Thus, in two deficient cell lines, induction activity was 39 and 35% that of control C12 cells before transfection with the PKA_c expression vector and was 56 and 39% of the activity observed in the control cells that had also been transfected with PKA_c. This is not the result that would be expected if the only difference that could affect *TH* gene expression in the control and the deficient cells was the relative lack of PKA in the deficient cells. Nonetheless, these results do support the proposal of Kim et al. (284) that PKA, by interacting with the CRE in the upstream region of the *TH* gene, plays two roles in the regulation of *TH* gene expression. First, the demonstration that *TH* expression is significantly decreased in PKA-deficient PC12 cells implicates PKA in the basal transcription and expression of the hydroxylase gene. Second, the loss of the ability of dibutyryl-cAMP to induce *TH* in the PKA-deficient cells supports the conclusion that PKA also plays a role in the long-term adaptation of *TH* to certain stimuli. Remember, as reviewed previously, that PKA has also been implicated in the acute activation of *TH*, which is mediated by the direct phosphorylation of the enzyme.

Since the only known metabolic function of PKA and CaM–PKII is the phosphorylation of proteins, the question that remained to be answered was the identity of the protein(s) whose phosphorylation by these kinases is so crucial to the CRE-dependent induction of TH. Before the nature of this protein was known, the evidence that transcriptional induction by cAMP is relatively rapid and is resistant to protein synthesis inhibitors, indicated that the candidate regulatory protein itself was not newly synthesized in response to cAMP, but rather that its phosphorylation was a key step in the response.

The identification of a likely candidate protein came from studies of the cAMP-mediated regulation of the transcription of the somatostatin gene in PC12 cells. Gonzalez and Montminy (285) found that stimulation of the expression of this gene by forskolin involves the phosphorylation at Ser^{133} of a 43-kDa nuclear CRE-binding protein, designated CREB. The evidence indicates that the phosphorylated CREB then combines with the CRE in the upstream region of the somatostatin gene and thereby induces transcription.

The CREB proved to be the link connecting certain transcriptional events that are stimulated by cAMP and those stimulated by membrane depolarization. Using the induction of the proto-oncogene c-fos in PC12 cell as a model, Sheng et al. (286) reported that membrane depolarization, known to activate transcription of this gene in a Ca^{2+}-dependent manner [reviewed in (286)], also stimulates the phosphorylation of CREB. Furthermore, the Ca^{2+} response element proved to be indistinguishable from the cAMP response element (CRE), suggesting that the common denominator involved in mediating the independent Ca^{2+} and cAMP signals that lead to activation of gene transcription is the phosphorylation of CREB. That these two signals are, in fact, independent, is supported by the observation that there is a marked synergism between the stimulatory effects of Ca^{2+} and cAMP agonists acting through the Ca^{2+} responsive element (286). Remember that a synergistic stimulation by veratridine-mediated depolarization and by cAMP analogs of TH gene expression in PC12 cells has also been observed (283). Also, in accord with the results of Sheng et al. (286) on regulation of expression of the somatastatic gene, the depolarization responsive element in the TH gene was mapped to the CRE region (283).

As noted previously, the evidence that the Ca^{2+} influx that accompanies membrane depolarization is crucial for the transcriptional response to this signal made it likely that a Ca^{2+}-dependent kinase is responsible for the phosphorylation of CREB. Furthermore, inhibitor studies indicated that the kinase involved in the depolarization-stimulated induction of TH is not PKC (283). In contrast to these negative results, induction of c-fos by depolarization was shown to be inhibited by calmodulin antagonists (287). In vitro studies have shown that CaM–PKI and CaM–PKII can both phosphorylate isolated CREB at Ser^{133} (288). The finding that the residue in CREB can also be phosphorylated by PKA (288, 289), provides a biological mechanism that can explain how Ca^{2+} and cAMP signaling pathways might converge to activate CREB and thereby gene expression.

Although many of the details of this integrated response involving phosphorylation of CREB came from studies of the regulation of other genes, there is little reason to doubt that a similar convergent mechanism underlies the transduction of the diverse signals that are involved in the regulation of the expression of the *TH* gene (283, 284).

IX. Human Tyrosine Hydroxylase and the Tyrosine Hydroxylase Gene

There is general consensus, supported by restriction endonucle-ase mapping, Southern blotting, and sequence analysis, that TH is encoded by a single gene that spans approximately 8 kb, and is com-posed of 13 introns. In lower mammals (i.e., non-primates) there is only one species of mRNA that generates a single form of the en-zyme. By contrast, in primates, and particularly in humans, the pri-mary mRNA transcript undergoes alternative splicing within intron 1 to generate multiple TH mRNA species [for a review, see (290)]. In monkeys, at least two species of TH mRNA have been found (291), whereas in humans, TH, whose gene has been localized to human chromosome region 11p15 (292), is encoded by four different mRNA species (293–297). These different mRNA species (type 1–4) are essentially indistinguishable except for a small area near the 5'-termini, where there is insertion of a 12, 81, or 93 bp sequences. Type 1 human mRNA, which has the highest sequence homology with rat TH, contains 1491 bp encoding 497 amino acids with a pre-dicted molecular weight of 55,973. Types 2 and 3 have an insertion of 12 and 81 bp, respectively, between the 90th and 91st nucleotides of type 1. Type 4 contains both the 12 and 81 bp (i.e., a total of 93 bp) nucleotide sequences, which are also inserted between position 90 and 91. These various insertions translate into enzyme species that are slightly larger than type 1; The number of amino acid resi-dues and the predicted molecular weights of types 2, 3, and 4 are 501 (M_r = 55,973), 524 (M_r = 58,080), and 528 (M_r = 58,521), respectively. Among the four types of human TH, types 1 and 2 are the most abundant and most widely distributed in the central nervous system as well as periphery. The two mRNA forms that have been detected in lower primates correspond to the human types 1 and 2 (291). On the other hand, types 3 and 4 are relatively rare, particularly in the brain, where they represent about 1% of total TH mRNA (298). In pheochromocytoma and adrenal medulla, the levels of types 3 and 4 are slightly higher, but they still only represent less than 5% of the TH mRNA. With respect to the distribution of the various forms of the enzyme in human adrenal medulla, there is considerable disagreement about the presence or absence of types 3 and 4. Thus, mRNA for type 4 was not detectable in a single gland (293), types 3 and 4 were missing from another (296), whereas low levels of both types were detected in a third gland (298). In a study that looked for TH protein, rather than TH mRNA, however, all four types of the enzyme were detected in six different postmortem samples of human adrenal medulla. Although it was not possible to quantitate each type, quantitation of the combined levels of type 3 plus type 4 and the combined levels of type 1 plus type 2 showed that types 3 and 4 represented 15–23% of the total TH protein present (299). The detection of type 4 TH in human adrenal tissue is of some significance because the failure to find it in this tissue together with the report that it is present in pheochromocytoma tumors lead to the conclusion that form 4 is characteristic of this tumor tissue (300).

To date, no remarkable differences in the catalytic properties of the four types of human TH have been detected. Expressed (and assayed with 6MPH$_4$ at pH 6.0) in COS cells, the relative homospecific activities, that is, activities normalized to the amount of TH immunoreactive protein, were reported to be type 1, 1.00; type 2, 0.40; type 3, 0.31; type 4, 0.26. The K_m values for 6MPH$_4$ were all in the fairly narrow range of about 180–250 μM; those for tyrosine ranged about 100–250 μM, with type 1 having the lowest value (105 μM) and type 4, the highest value (253 μM) (301).

In agreement with the results in COS cells, the specific hydroxylase activities of types 2, 3, and 4 expressed in *Xenopus* oocytes were found to be lower than that of type 1 (300). The relative activities were type 1 (hTH1), 1.00; type 2 (hTH2), 0.35; type 3 (hTH3), 0.58; and type 4 (hTH4), 0.59. These relative values are similar, but not identical, to those found in COS cells, the main difference being that in oocytes, the specific activities of hTH3 and hTH4 were higher than that of hTH2, whereas in COS cells, the specific activity of type 2 was higher than that of types 3 and 4.

Horellou et al. (300), noting that types 2, 3, and 4 have an extra arginine residue in the N-terminal region, proposed that this extra positive charge could account for the lower specific activity of these forms. This proposal is in accord with the earlier discussion of how the regulatory and catalytic domains of TH interact to control enzyme activity and predicts that PKA-mediated phosphorylation of types 2, 3, and 4 might activate these species to a greater extent than that seen with type 1. From a comparison of the effect of PKA-mediated phosphorylation of essentially pure hTH1 and hTH2 (expressed in *E. coli*), however, there was no indication that hTH2 was activated to a greater extent than hTH1, although more phosphate was incorporated into hTH1 ($\sim$0.75 mol/mol of subunit) than into hTH2 ($\sim$0.5 mol/mol subunit) (197). In fact, in agreement with the findings with recombinant rat TH isolated from *E. coli* (198, 200), neither type was substantially activated by PKA-mediated phosphorylation. Thus, V_{max} (measured with BH_4 at pH 7.0) was unchanged, although the K_m of both forms for BH_4 was modestly ($<$ twofold) decreased and the K_i for dopamine was increased. Only with this last parameter was there any indication that the phosphorylation-induced change might lead to a greater activation of hTH2 than of hTH1; that is, the K_i for dopamine was increased more for hTH2 (from 0.3 to 1.2 μM) than for hTH1 (from 0.5 to 1.5 μM).

The finding of four different forms of human TH has fostered the widely held view that this diversity has functional significance. LeBourdelles et al. (197), for instance, speculated that the alternative splicing of pre-mRNA, which underlies the formation of these different forms, may regulate long-term TH activity and, more specifically, that neuronal activity may modulate the splicing process. Implicit in this view is the notion that these different forms of the enzyme differ significantly in their properties.

As already noted, hTH2, hTH3, and hTH4 all have lower specific activities than hTH1 and, going along with the lower catalytic activity, they also generally have somewhat less favorable K_m values for tyrosine and for the pterin cofactor. Also, as mentioned above, the effects on hTH1 and hTH2 of phosphorylation by PKA are essentially the same, despite the fact that more phosphate is incorporated into hTH1 than into hTH2.

Perhaps of greater potential functional significance, as has been pointed out by Grima et al. (293), is the fact that the addition of four extra amino acids (Val-Arg-Gly-Gln) in the N-terminal region of hTH2, confers on the adjacent serine residue (residue 31 in hTH1) the potential for being phosphorylated by CaM–PKII. That hTH2 does indeed contain an additional phosphorylation site was demonstrated by the finding that CaM–PKII catalyzes the incorporation of about 2 mol of phosphate per mol of subunit of hTH2, whereas only about 1 mol of phosphate per mol of hTH1 subunit is incorporated. Furthermore, the additional site in hTH2 phosphorylated by CaM–PKII was identified as Ser[31]. In accord with results with rat TH, it was also shown that in both hTH1 and hTH2, Ser[19] is phosphorylated only by CaM–PKII, whereas Ser[40] is phosphorylated by both PKA and CaM–PKII (197). Since hTH2 has three sites capable of being phosphorylated by CaM–PKII (i.e., Ser[19], Ser[31], and Ser[40]), it is not clear why only 2 mol of phosphate per mol of hTH2 subunit were incorporated into this isoform. Phosphorylation of hTH2 by CaM–PKII does not change either V_{max} or the K_m for BH$_4$, both values measured at pH 7.0. Surprisingly, however, it did increase by about twofold the K_i for dopamine (197), an indication that the section of the N-terminal region that is altered in hTH2 by the addition of the four extra amino acids plays a role in binding dopamine to the enzyme.

The consensus view that the four different isoforms of human TH are of some functional significance is certainly a reasonable one. It is also reasonable, however, that this view be supported by evidence that the different forms differ significantly in their properties. So far, however, these differences, some of which have been reviewed above, do not provide even a hazy picture of what the functional significance of the various forms might be.

X. Structural Studies on Tyrosine Hydroxylase

In Section VI(C), a structural aspect of TH that is related to activation of the enzyme by phosphorylation was discussed, specifically, the location of the serine residues that are phosphorylated by various kinases. In this section, some of the more global structural features of the hydroxylase will be considered.

Most of the structurally relevant data have come from studies of limited proteolysis or mutagenesis of the enzyme. As mentioned earlier, partial digestion with trypsin (79, 91) or chymotrypsin (16) showed that TH could be hydrolyzed to a catalytically active species with an apparent molecular weight of 34 kDa (80, 81, 89). These results showed that, like PAH (82), TH contains a protease-resistant core and that the catalytic domain is located in this core. Furthermore, the finding that the 34-kDa fragment generated by chymotrypsin digestion of the hydroxylase could no longer be activated by phosphorylation (81), indicated that the regulatory domain had been removed by the action of the protease and that this fragment also lost some structural feature that is critical for oligomerization of the protein.

A further analysis of the functional organization of TH showed that partial digestion of the enzyme with trypsin removed a 17-kDa N-terminal fragment and a 5-kDa C-terminal fragment from the 56-kDa monomer (302, 303). [The 56-kDa value is based on the deduced amino acid sequence (84); Abate et al. (302, 303), used a value of 60 kDa based on results of sodium dodecylsulfate (SDS)-polyacrylamide electrophoresis.] These studies confirmed the finding (91) that the resulting 34-kDa species is in an activated state and that it lacks all four phosphorylation sites located in the N-terminus (303). It was also shown that the proteolyzed hydroxylase has many of the properties of the activated native enzyme, including a lower K_m for 6MPH$_4$, in accord with the earlier finding that it had a lower K_m for DMPH$_4$ (91), and a higher pH optimum, properties that are analogous to those of the phosphorylated native enzyme (85). In contrast to the effect of phosphorylation, however, proteolysis also decreased the K_m for the substrate tyrosine as originally reported by Kuczenski (91), and even more dramatically decreased the K_m for phenylalanine (303). These findings are consistent with the model that the N-terminus of TH, like that of related aromatic amino acid hydroxylases (304), is the regulatory domain that directs cofactor binding and substrate specificity.

Additional insight into the structure of TH came from studies of deletion mutagenesis of the cloned rat PC12 enzyme (305). The results are depicted schematically in Figure 4. As can be seen, deletion of the N-terminal 132 amino acids (ΔA), doubled hydroxylase activity. The deletion of 157 amino acids (ΔB) from the N-terminal region further increased the activity relative to the wild type (wt), whereas the deletion of up to 184 N-terminal residues (ΔC) totally destroyed activity. Similarly, removal of the carboxy-terminal 43 amino acids

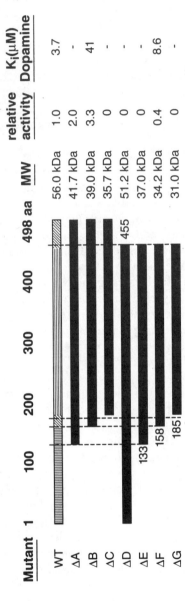

Figure 4. Schematic representation of TH deletion mutants. The seven mutant products (ΔA–G) and the wild type (WT) are shown. The regulatory domain is indicated by vertical lines and the horizontal lines depict the region that is predominantly catalytic. The TH activities were measured in crude extracts and are relative to that of the wild type (0.07 μmol/min/mg of protein) (305).

generated enzyme species that were inactive (ΔD, ΔE, and ΔG). The only exception was mutant ΔF, which lacked both 43 carboxyl-terminal amino acids and 157 amino acids from the N-terminus. This mutant had about 40% of the wild-type activity.

These findings further defined the boundaries of the catalytic domain, narrowly circumscribing its amino end to the region between residues 158 and 184. The carboxyl end of the catalytic domain, however, is less clearly defined. Although the finding that removal of the 43 carboxyl terminal amino acids eliminated hydroxylase activity (e.g., in mutants ΔD, ΔE, and ΔG), raised the possibility that this 43 amino acid deletion had entered the right edge of the catalytic domain, such a possibility is not supported by the finding that mutant ΔF, generated by the combined deletion of 157 residues from the N-terminus together with the deletion of 43 carboxyl terminal residues, has substantial, albeit less than wild-type, hydroxylase activity. This last result indicated that the catalytic domain is still largely functional in a species that lacks 43 amino acids from the carboxyl terminal end of the molecule. Rather than being part of the catalytic domain, therefore, this segment of the carboxyl terminal region may be involved in the correct folding of the catalytic site, or, as suggested by Liu and Vrana (306), in maintaining the tetrameric conformation of the enzyme.

The finding that the double deletion mutant ΔF has lower activity than the wild-type enzyme was unexpected since this construct most closely resembles the proteolytically activated 34-kDa species studied by Joh and co-workers (302, 303). A strict comparison between these two truncated forms of the enzyme, however, is difficult because the precise localization of one of the bonds cleaved by trypsin to generate the activated 34-kDa species is unknown. Thus, sequence analysis of the N-terminal region of this fragment identified the hydrolyzed bond as the one between Arg^{157} and Ser^{158}. The bond cleaved in the C-terminal region, however, was not identified, although Abate et al. (302) suggested the arginine at position 455 [designated Arg^{454} by Abate et al. (302)] as a likely candidate because cleavage at this site would generate a fragment of the correct molecular weight (i.e., $\sim$34 kDa). Indeed, following this suggestion, this site in the C-terminal region was selected in the construction of mutant ΔF (305). Although this might have seemed like a likely cleavage site, the fact that the amino acid residue at position 456 is a proline

(84), when taken together with the fact that trypsin does not cleave the Arg-Pro bond (307), makes this assignment of one of the trypsin cleavage sites unlikely. There are at least five other lysine or arginine residues within about 12 amino acid residues of Arg[455] and most of them are reasonable candidates for the trypsin cleavage site. Not until the C-terminal residue in the proteolytically activated 34-kDa species has been identified can a strict comparison be made between the catalytic properties of this species and the corresponding one generated by mutagenesis. If truncation at one of these other potential trypsin-sensitive sites should generate an activated species of the hydroxylase, it would prove that this narrow sequence in the C-terminal region of the molecule is a critical determinant of the activity of the enzyme.

In addition to the differences in catalytic properties just noted, there are other differences between the truncated species generated by trypsin proteolysis and by deletion mutagenesis. For example, proteolytic removal of only a 4-kDa fragment from the N-terminus markedly decreased the K_m for 6MPH$_4$ (303), whereas mutagenic deletion of enough of the N-terminal region to activate the hydroxylase two- to threefold (mutants ΔA and ΔB, Fig. 4), did not decrease the K_m for BH$_4$ (305). These truncated mutants also differ from their proteolytically derived counterparts in their relative affinities for tyrosine and phenylalanine. Thus, in agreement with the results of Abate and Joh (303), activated mutants ΔA and ΔB have lower K_m values for tyrosine (305). However, whereas limited proteolysis was found to cause a disproportunately large decrease in the K_m for the alternate substrates phenylalanine and tryptophan (e.g., the K_m for tyrosine was decreased to about one-half of the wild-type value compared to a 20-fold decrease in the K_m for phenylalanine), the K_m values for both tyrosine and phenylalanine were each decreased by the same amount (~50%) in the deletion mutants ΔA and ΔB (305). In the double deletion mutant ΔF, by contrast, most closely resembling the proteolytically activated 34-kDa species, the K_m for tyrosine was the same as that of the wild type and the K_m for phenylalanine was only slightly decreased compared to the wild-type value. Based on their finding that limited proteolysis yielding the 34-kDa species causes a preferential decrease in the K_m for phenylalanine and tryptophan, together with the observation that digestion of the C-terminal region alone did not decrease the K_m for the amino acid substrates, Abate and Joh (303) concluded that the N-terminus of TH restricts the substrate specificity of the enzyme, and that its removal results in a broadening of the amino acid substrate specificity. Although the results with the deletion mutants are coherent with the conclusion that the N-terminus plays a role in substrate binding, these results do not support the conclusion that this region can intrinsically direct substrate *specificity*.

These discrepancies suggest that there are one or more structural differences between the cloned enzyme isolated from *E. coli* and the native enzyme and that these differences may account for some of the divergent results obtained with proteolytic truncation and deletion mutagenesis of the enzyme. Specifically, the results raise the possibility that some eukaryotic posttranslational modification, which is missing in *E. coli,* may be critical for the expression of the enzyme's substrate specificity.

One such modification is suggested by the finding, reviewed earlier, that the native enzyme contains small amounts of tightly bound catecholamines, which are coordinated to iron in the enzyme (201). Several lines of evidence, going all the way back to the kinetic data showing that catecholamines are potent inhibitors of the enzyme (12), indicate that bound catecholamines keep the hydroxylase in a state of low activity. Thus, as noted earlier, Haavik et al. (201) showed that phosphorylation of the purified enzyme accelerates the dissociation of the bound catecholamines, a process that probably accounts for at least part of the phosphorylation-mediated activation of the enzyme.

The absence of bound catecholamines in the recombinant enzyme isolated from *E. coli* may be the crucial structural feature that underlies the different results obtained by partial proteolysis and by deletion mutagenesis. One of the reasons for these discrepant results may simply be that the cloned enzyme from *E. coli* already has many of the attributes of the activated enzyme, including those of the enzyme activated by limited proteolysis. The K_m for BH$_4$ (56 μM at pH 6) and the pH optimum (pH 6.3–7.0) of the cloned enzyme (199), are more characteristic of the activated enzyme [activation by either phosphatidyl-L-serine (66) or phosphorylation (51)] than of the native enzyme. Moreover, phosphorylation of the cloned enzyme only modestly decreases the K_m for BH$_4$ and increases the pH optimum (199). In view of these properties of the cloned enzyme, it may not be surprising that deletion of up to 157 amino acid residues from the inhibitory N-terminus of the cloned enzyme, which removes all of the phosphorylation sites, affects neither the K_m for BH_4 nor the pH optimum (305). By contrast, proteolytic removal of part of the N-terminus from the native enzyme, presumably in a state of low activity because of its bound catecholamine, decreases the K_m for the pterin cofactor and increases the pH optimum (303). Another finding that points to the affinity of the hydroxylase for catecholamine as one of the factors that can determine the properties of the truncated enzyme is that mutations that disrupt the N-terminus have a dramatic effect on dopamine binding. Thus, the K_i for dopamine inhibition was found to be increased nearly 10-fold when 157 amino acid residues were deleted from the N-terminus of the recombinant enzyme (Fig. 4, mutant ΔB). This finding is in accord with the results of a study of limited proteolysis of the bovine adrenal hydroxylase by calpain, which showed that removal of the first 31 residues from the N-terminus (counting from the start methionine), which led to a slight activation, increased the K_i for dopamine by 16-fold (from 4.4 to 72 μM), and that for norepinephrine by sevenfold (from 2.3 to 17 μM) (308).

Although the observation that deletions in both ends of the TH molecule can affect dopamine binding was unexpected, it is coherent with the idea that bound dopamine plays an important role in the regulation of the activity of the enzyme. In fact, these results indicate that there is a reciprocal relationship between the K_i values of the various mutants for dopamine and their catalytic activity. As seen in Figure 4, deletion of 157 residues from the N-terminus (mutant ΔB), activates the enzyme 3.3-fold and this change is accompanied by an increase in the K_i for dopamine by more than an order of magnitude. The further deletion of part of the C-terminus (mutant ΔF) decreases the activity by about an order of magnitude (compared to mutant ΔB) and also decreases the K_i for dopamine by a factor of 5. Clearly, some variable other than the affinity of dopamine must account for the fact that mutant ΔF, despite its having a higher K_i for dopamine, has less activity than the wild-type enzyme. It should be noted that this effect on dopamine binding of deletion of part of the C-terminal region provides a second example of the interaction between this region and the N-terminal regulatory domain of the enzyme. The other evidence for such an interaction can be seen in the ability of the deletion of 157 amino acids from the N-terminus to partially reverse the deleterious effect of deletion of 44 amino acids from the C-terminus, that is, mutant ΔD, with only the C-terminal deletion is devoid of activity, whereas the additional deletion of the N-terminal region, as in mutant ΔF, restores 50% of the wild-type activity (see Fig. 4).

Deletion mutagenesis studies of TH have also put into sharp focus a difference between the effects on the enzyme of dopamine and dopa. Early studies on the inhibition by various catechols did not distinguish between the effects of these two metabolites. Rather, inhibition by both catecholamines and dopa was reported to be competitive with the tetrahydropterin cofactor (308, 309). An indication that this generalization is incorrect was provided by the observation that inhibition by dopa of the bovine adrenal enzyme is not competitive with either BH_4 or $DMPH_4$ (15). In line with these results, it was reported that whereas inhibition of the bovine brain enzyme by dopamine is competitive with BH_4, inhibition by dopa is noncompetitive (120).

Even more dramatic differences between the effects of dopamine and dopa were observed with the cloned hydroxylase and mutants derived from it. In contrast to dopamine, dopa does not significantly increase the K_m for BH_4. Moreover, the K_i for dopa is 17-fold higher than that for dopamine and none of the deletion mutants studied appeared to substantially change that K_i value (305). These results suggest that dopa does not play the same crucial role as catecholamines in the regulation of the enzyme.

The absence of bound catecholamines in the cloned enzyme and the finding of a rough inverse correlation in the deletion mutants between the tightness of dopamine binding and the state of activation of the enzyme offers a possible explanation for the divergent finding that the 34-kDa species generated by trypsin proteolysis is activated, whereas the analogous species produced by deletion mutagenesis of the already activated cloned enzyme is not and, in fact, has somewhat less activity than the wild-type enzyme (305).

It is possible that the explanation for this discrepancy, as discussed earlier, can be traced to the uncertainty about which bond in the C-terminal region was actually cleaved by trypsin to generate the 34-kDa species and the consequent improbability that the 34-kDa species produced by mutagenesis is identical to one derived by proteolysis.

An alternate explanation, however, assumes that limited proteolysis of the enzyme of the type under consideration has two opposite effects on hydroxylase activity. One of these effects is a large activation due to removal of the inhibitory N-terminal region. This activation would probably be enhanced by the decreased affinity for bound catecholamines resulting from removal of the inhibitory N-terminal region, as observed in the N-terminal deletion mutants (305). It should be emphasized, incidentally, that the activation observed by deletion of up to 157 amino acids from the N-terminus of the cloned enzyme demonstrated for the first time that the N-terminal region exerts an inhibitory effect on the enzyme even in the absence of bound catecholamines. In addition to this activating effect of removal of the N-terminus, however, proteolysis has a second effect, that is, a decrease in intrinsic hydroxylase activity consequent to removal of part of the C-terminal region. That this can occur is demonstrated by the finding that deletion of 43 amino acids from the C-terminus can totally destroy hydroxylase activity (see Fig. 4) (305). With the native enzyme-containing bound catecholamines, the net effect of these two opposite changes is activation, as observed by Abate and Joh (303). With the cloned enzyme, however, already in a partially activated state because of the absence of bound catecholamines, the net effect of truncation of both N- and C-terminal regions might be different. Deletion of the inhibitory part of the N-terminus would still be expected to result in activation, as observed in mutant ΔB, but because of the absence of bound catecholamines, this activation would probably be attenuated compared to the effect on the native enzyme. It is not unlikely, therefore, that the net effect of a smaller activation due to deletion of the N-terminal region and the unattenuated adverse effect of deletion of the C-terminal region would be a decrease in catalytic activity compared to the parent wild-type cloned enzyme, as observed by Ribeiro et al. (305).

These studies of the limited proteolysis and deletion mutagenesis of TH have provided important insights into the structural features of the enzyme that underlie its regulation. Figure 5 shows in schematic form a model for the structure of the enzyme that incorporates some of the results of these studies. The present model expands on

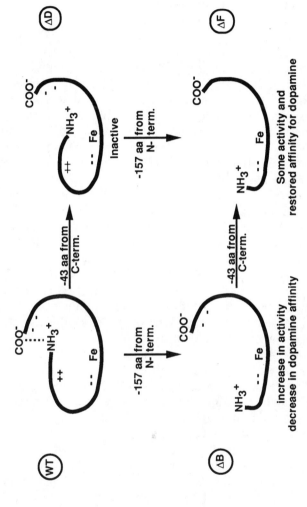

Figure 5. Model for TH depicting schematically the effects of truncation on the enzyme's structure and activity (aa, amino acids). [Adapted from (305).]

previous ones (130, 303), which, in turn, are all versions of the original "internal inhibitor" model for PAH that was proposed to explain the activation of that enzyme by limited proteolysis and exposure to certain phospholipids (82). The present model is also comparable to the more recent one for PAH (310). Before discussing the details of the present model, note that the low activity states of the enzyme are depicted as "tight" or closed structures in which the Fe at the active center is relatively inaccessible to substrates, whereas the activated states are shown as "relaxed" or open structures allowing free access of reactants to the active center. It should be emphasized, however, that accessibility to the Fe center is used here as a symbolic, rather than a literal, depiction of low- and high-activity states.

The model incorporates the notion that TH is organized into a regulatory domain located within the first 157 residues of the N-terminus, and a C-terminal catalytic region, which resides between residues 158 and 184 on the amino end, and up to position 455 on the carboxyl end. The first 157 amino acids in the N-terminus of TH have an overall inhibitory effect on the enzyme, such that the removal of this region results in an increase in activity, a decrease in the K_m for the substrate, and a dramatic decrease in the affinity for dopamine and sensitivity to dopamine inhibition (see Fig. 4). As depicted in Figure 5, deletion of the inhibitory N-terminal region (157 amino acids) leads to a relaxed, activated structure (mutant ΔB), whereas deletion of part of the C-terminal (mutant ΔD) leads to a collapsed, inactive species. One of the more difficult structure–activity relationships to explain is how deletion of 157 amino acids from the N-terminus can restore significant activity to the inactive species generated by deletion of 43 amino acids from the C-terminal. An attempt to explain this paradoxical effect is shown in Figure 5. This deletion from the C-terminal, perhaps because it disrupts the balance of attractive and repulsive charges, leads to the collapse of the structure, converting it into a tight, inactive configuration (mutant ΔD). The further deletion of a critical region of the N-terminus to generate the double deletion mutant (ΔF), may be able to approximate the original balance of charges, leading to a more relaxed structure with significant catalytic activity.

With respect to this postulated role of electrostatic charges as a determinant of the three-dimensional (3-D) structure of both the native and the activated enzyme, a role which, incidentally, is not meant to preclude an important contribution of hydrophobic interactions, this general picture is in accord with a similar one for PAH that grew out of the original "internal inhibition" model for that enzyme. The demonstration that PAH can be activated by phosphorylation (131) implied that the inhibition was due to the electrostatic interaction of the positively charged regulatory domain with a negatively charged region in or near the catalytic domain, and that phosphorylation of the regulatory domain activated the enzyme by disrupting this interaction, displacing the regulatory domain.

The observation that TH can be activated by heparin (25) and phosphatidyl-L-serine (66), which are probably specific examples of a more general activation of the enzyme by polyanions (67), indicated that TH and PAH are constructed along similar lines, with a similar structural motif involving the interaction of positively and negatively charged areas of the protein.

This notion was further strengthened by the finding that like PAH, TH can also be activated by phosphorylation. In a further development of this idea, Katz et al. (67), as noted earlier, postulated that activation by polyanions involved their electrostatic interaction with regions of high positive density on the enzyme. Strong support for this notion came from the findings that the activated 34-kDa species generated by limited proteolysis can no longer be activated by either phosphatidyl-L-serine or by phosphorylation (81), or by polyanions (89). These results suggested that the regions removed by proteolysis contained the positively charged anionic binding site(s). The results of Abate and Joh (303) and Ribeiro et al. (305) can be used to probe the validity of these early suggestions implicating a positively charged area in activation of the enzyme. Since both of these studies have defined which regions of the native enzyme are removed to generate the 34-kDa species, it should be possible to determine the charge characteristics of the deleted sequences. A comparison of the mutation deletion data of Ribeiro et al. (305), which permit a precise localization of the boundaries of the deletions, with the sequence data (84), indicates that at pH 7.0 the 157 amino acid region that is deleted from the N-terminus is indeed a region with a net positive charge. The 43 amino acids region deleted from the C-terminus is a region with an even larger net negative charge. Acknowledging the limits of any attempt to extrapolate linear sequence data to a 3-D structure, these comparisons are, at least, coherent with the notion that the regulatory domain, with a net positive charge, can interact with a negatively charged region at the edge of the catalytic domain to maintain a "tight," configuration, as depicted in Figure 5. In addition, the sequence of 157 amino acids from the N-terminus with its net positive charge could be the site of interaction with activating polyanions (67, 89).

XI. Mechanism of Action

Studies on the mechanism of action of TH have not progressed as far as those on PAH, mainly because until recently, large amounts of the pure enzyme were not available.

In the absence of any evidence to the contrary, it is reasonable to assume that the mechanism of action of TH (as well as that of TPH) is the same as that of PAH. Even before any mechanism studies had been carried out, there could be little doubt about the correctness of this assumption. It is strongly supported by both structural and biochemical evidence. With respect to the structural evidence, there is a high degree of sequence homology between PAH and TH at both the nucleic acid and amino acid levels (311). Moreover, it is known that TH, like PAH (312) is a nonheme iron protein (16, 81). Some of the biochemical evidence pointing to a common mechanism was the early finding that with catalytic amounts of 6MPH$_4$, the TH-catalyzed reaction is dependent on the presence of DHPR and a reduced pyridine nucleotide (10), which provided powerful evidence that quinonoid dihydropterin is the ultimate pterin product formed during the hydroxylation reaction. A common mechanism was also supported by the demonstration that with BH$_4$ as the coenzyme, the stoichiometry of the TH-catalyzed reaction is the same as that catalyzed by PAH [see Eq. (1)] (16).

In view of the many indications that the mechanisms of TH and PAH are the same, it was anticipated that, just as had been shown with PAH (17, 313), the 4a-hydroxytetrahydropterin would also be the primary pterin product formed during the TH-catalyzed reaction. Studies carried out with TH isolated from both cultured PC12 cells (75) and from bovine adrenal medulla (72) have shown that the 4a-carbinolamine is indeed the pterin product formed during the TH-catalyzed reaction. In the latter study, formation of this product was also demonstrated during the TH-catalyzed hydroxylation of phenylalanine. As previously shown with PAH (17, 313), during early times, equal amounts of the carbinolamine and hydroxylated amino acid product are formed during the TH-catalyzed reaction (72, 75). The overall TH-catalyzed reaction, coupled to DHPR-catalyzed regeneration of the tetrahydropterin is shown in Figure 6.

Just as with the PAH-catalyzed hydroxylating reaction, where it has been shown that the oxygen in the newly formed tyrosine is derived from molecular oxygen (314), it has been shown that during the hydroxylation of tyrosine catalyzed by adrenal TH, the extra oxygen in the product, dopa, is derived from molecular oxygen (315). It was also shown that there was virtually no migration of the oxygen from the 4 position of tyrosine to the 3 position of dopa, that is, atmospheric oxygen is the source of all of the oxygen in the 3 position of dopa. Furthermore, in experiments carried out with phenylalanine labeled with tritium in the 4 position, it was found that the newly synthesized dopa contained 42% of the tritium that was originally in the 4 position of the phenylalanine, an indication that during the

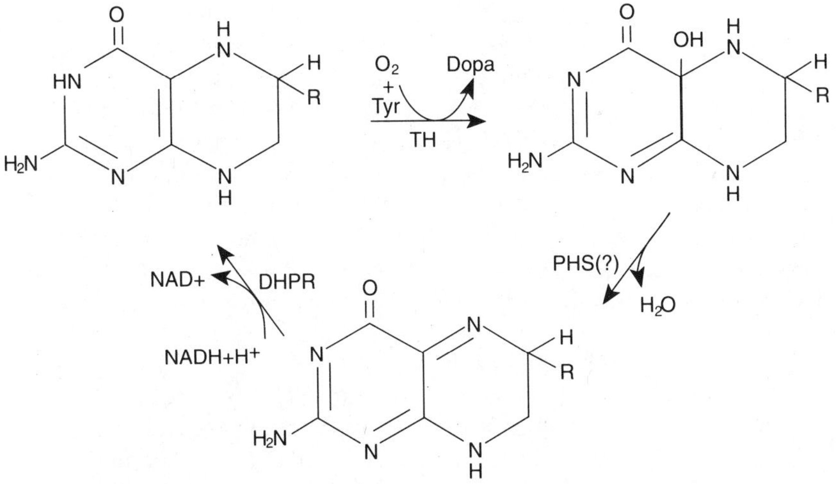

Figure 6.   Scheme for the enzymatic conversion of tyrosine to dopa catalyzed by TH. The "PHS" is a PAH stimulator, also called pterin carbinolamine dehydratase. (The question mark is included because stimulation of TH by the dehydratase has not yet been demonstrated.)

hydroxylation of phenylalanine catalyzed by TH, a shift of the tritium to the 3 position of the tyrosine occurs. Although migration of the hydroxyl group does not occur, migration of tritium from the 4 to the 3 position does take place during the TH-catalyzed hydroxylation of (4-^{3}H)phenylalanine. On the other hand, the hydroxylation of either (3,5-^{3}H)phenylalanine (315) or (3,5-^{3}H)tyrosine (12) leads to the loss of about one-half of the tritium. These results showed that the first hydroxylation of (4-^{3}H)phenylalanine in the 4 position catalyzed by TH leads to the migration of the tritium to the 3 and 5 position (just as it does during the PAH catalysis of this same reaction) and that the second hydroxylation of the resulting (3,5-^{3}H)tyrosine in either the 3 or 5 position is accompanied by the loss of the ^{3}H that originally occupied that position. Based on these results, it has been postulated that the hydroxylation of phenylalanine by TH proceeds by the same mechanism as the PAH-catalyzed oxidation of phenylalanine, that is, a protonated epoxide species of either phenylalanine or tyrosine (involving the para or meta positions of the amino acid) is formed as an intermediate in the reaction (315).

A kinetic analysis of the TH-catalyzed reaction carried out with partially purified soluble adrenal enzyme in the presence of DMPH$_4$ led to the conclusion that the reaction is of the Ping Pong type, involving a reduced form of the enzyme as an intermediate (309). This conclusion was based, in part, on the observation that double reciprocal plots of initial rates versus tyrosine concentrations at different concentrations of DMPH$_4$, gave a series of parallel lines.

Subsequently, Joh et al. (316) carried out kinetic studies with the solubilized adrenal enzyme and concluded that the mechanism did not involve a reduced enzyme intermediate but rather a quaternary complex. Since their findings with the trypsin-solubilized enzyme were different from those reported for the soluble enzyme (309), these authors proposed that the mechanism of action of the soluble and solubilized enzymes was different.

Results obtained by Shiman and Kaufman (15) with both the particulate and the highly purified, solubilized bovine adrenal enzyme, failed to confirm the original findings of Ikeda et al. (309) and also made unlikely the possibility that the solubilized and particulate enzymes differ in their mode of action. Shiman and Kaufman (15) found that for both forms of the enzyme, double reciprocal plots of initial rates versus either DMPH$_4$ or BH$_4$ concentrations at several different tyrosine concentrations gave a series of intersecting lines, indicating that the mechanism is not of the Ping Pong type, as described by Ikeda et al. (309), but rather of the sequential type in which tyrosine must bind to the enzyme before the dihydropterin product is released. Based on their kinetic analysis, Shiman and Kaufman concluded that the TH reaction proceeds through a quaternary enzyme–substrate complex involving the hydroxylase-tetrahydropterin-tyrosine-oxygen and that a reduced form of the enzyme is not an intermediate in the reaction.

With partially purified soluble (in contrast to "solubilized") TH from bovine adrenal medulla, Oka et al. (65) reported that varying BH_4 and oxygen or BH_4 and tyrosine gave an intersecting pattern in Lineweaver–Burk plots of the data, whereas varying tyrosine and oxygen gave parallel lines. They also confirmed the observation (16) that in the presence of BH_4, concentrations of tyrosine greater than 50 μM inhibit hydroxylase activity. Oka et al. (65) drew no conclusions about the kinetic mechanism from their data.

In contrast to the results reported by Oka et al. (65) but in confirmation of those found by Shiman and Kaufman [reported in (15)], Bullard and Capson (120), using partially purified bovine striatal TH, observed intersecting line patterns with all three substrate pairs. On the basis of these results and those from product inhibition studies, they also concluded that the mechanism is sequential with the formation of a quaternary complex occurring before any irreversible catalytic step or product release. Furthermore, they concluded that either tyrosine or BH_4 could bind to the enzyme first but that oxygen probably binds after BH_4. These workers also replicated the earlier original observation that inhibition by dopa is noncompetitive with respect to BH_4.

This kinetic mechanism was consolidated and further elaborated by Fitzpatrick (317), who, based on studies with $6MPH_4$, also concluded that the mechanism involves formation of a quaternary complex before any irreversible step occurs. The data indicated that formation of this complex results from the ordered binding of substrates, the order being $6MPH_4$, oxygen, and tyrosine. In an earlier study, the observations that high concentrations of $(6R)$-BH_4 (88 μM) are unable to abolish the inhibition by excess tyrosine led to the conclusion that formation of a dead-end enzyme–tyrosine complex could not explain the inhibition by excess tyrosine (318). More recently, however, the opposite conclusion was reached. It was found that, in contrast with the results obtained with BH_4, tyrosine is a competitive inhibitor versus $6MPH_4$, leading to the conclusion that a dead-end enzyme–tyrosine complex is formed (317). It was also reported that there is no burst of dopa release during the first turnover of the enzyme, an indication that product release is not rate limiting. The finding that there is no primary kinetic isotope effect on either V_{max}/K_{tyr} or on V_{max} with $(3,5-{}^2H_2)$tyrosine as the substrate also showed that cleavage of the carbon–hydrogen bond in tyrosine is not rate limiting. From these kinetic results, and from the lack of inhibition of TH by possible transition state analog (73), it was concluded that formation of the hydroxylating intermediate is the slow step in the tyrosine-catalyzed reaction.

It should be noted that the conclusion that cleavage of the carbon–hydrogen is not rate limiting, is coherent with the one reached from studies of the mechanism of PAH carried out with ring-deuterated phenylalanine. This mechanism showed no isotope effect with BH_4 on V_{max} or V_{max}/K_m and only a small effect on V_{max} with $6MPH_4$ (319). From the failure to detect an isotope effect in the presence of BH_4, it was postulated that in the presence of the natural coenzyme, the rate-limiting step in the PAH-catalyzed hydroxylation of phenylalanine is the formation of the tetrahydropterin hydroperoxide, the putative hydroxylating species (319).

Based on his results, Fitzpatrick (317) proposed a kinetic mechanism for TH, depicted in Figure 7, which shows the ordered addition of substrates to the enzyme, as well as the formation of a dead-end enzyme–tyrosine complex to account for inhibition by excess tyrosine.

The scheme showing the enzyme combining with the tetrahydropterin prior to tyrosine fits in well with the observation that the K_m for the amino acid substrate varies markedly with the nature of the pterin coenzyme used, being much lower in the presence of BH_4 than either $6MPH_4$ or $DMPH_4$ (48, 64). The K_m of the other substrate, oxygen, also varies with the tetrahydropterin; like the K_m for tyrosine, it is lower with BH_4 than with $DMPH_4$ (320). This finding again can be explained by a sequence in which oxygen combines with the enzyme *after* the tetrahydropterin.

Although the proposal of a dead-end tyrosine–enzyme complex is consistent with the observed inhibition by excess tyrosine (16), it

Figure 7. Scheme illustrating the kinetic mechanism for TH. The XH_4 stands for the tetrahydropterin cofactor. [After Fitzpatrick (317).]

does not readily explain the finding that inhibition is more pronounced in the presence of BH_4 than it is with synthetic model tetrahydropterins (16). Indeed, the scheme suggests that with BH_4, which has a much lower K_m than $6MPH_4$ and a comparable V_{max} value (see Table II), inhibition by excess tyrosine would be less pronounced than it is with $6MPH_4$. In addition, the scheme is clearly an oversimplification since it does not try to account for the finding that excess oxygen also inhibits and, like the situation with tyrosine, inhibition is more severe in the presence of BH_4 than with a model coenzyme like $DMPH_4$ (320). Moreover, as discussed earlier (320), the degree of inhibition is proportional to the K_m for oxygen, being greater where the K_m for oxygen is small, as it is with BH_4, than it is when the K_m for oxygen is large, as it is with $DMPH_4$. To explain this correlation, it has been proposed that inhibition by excess oxygen may be due to the binding of a second molecule of oxygen to the substrate-binding site (320).

Despite the fact that the available evidence indicates that the adrenal and brain enzymes appear to proceed through the same type of kinetic mechanism, there do seem to be some differences. Perhaps the most noteworthy of these is the finding that with brain enzyme, increasing concentrations of either BH_4 or tyrosine decrease the enzyme's apparent affinity for the other compound (120). This reciprocal relationship is in sharp contrast to the findings reported with the adrenal enzyme where the K_m of neither $DMPH_4$ nor BH_4 were affected by increasing concentrations of tyrosine (15). Another apparent distinction was the failure to observe inhibition by excess tyrosine with the striatal enzyme (120). With respect to this observation, however, it is not a reflection of a difference in properties between the adrenal and brain enzymes, as implied by Bullard and Capson (120), since inhibition by tyrosine has been reported for the adrenal enzyme (16), as well as for brain enzyme, inhibition of the latter enzyme being observed both *in vitro* (66, 71) and *in vivo* (321). Rather, the explanation for this apparent discrepancy is simply that the highest concentration of tyrosine used by Bullard and Capson (120) was 25 μM, which is below the 50 μM tyrosine needed to see inhibition with both adrenal and brain TH (16, 71, 76).

A significant breakthrough in our understanding of the mechanism of TH seemed to be heralded by the report that the pure enzyme from PC12 cells can hydroxylate tyrosine when hydrogen peroxide

(H_2O_2) was substituted for a tetrahydropterin coenzyme (75). In accord with previous reports that TH is inhibited by high levels of H_2O_2 (16), it was found that a 100-fold molar excess of this agent completely inactivated the enzyme. In contrast, incubation with only a 10-fold excess of H_2O_2 in the absence of any tetrahydropterin resulted in the formation of slightly more hydroxylated product than the amount of enzyme present (0.61 μM of product formed with 0.5 μM of TH). The authors postulated that an Fe^{2+}—H_2O_2 complex or a ferryl species derived from the interaction of Fe^{2+} and H_2O_2 might be the hydroxylating species.

Attempts to replicate this observation, however, were completely unsuccessful: No hydroxylation of tyrosine was detected with the 10-fold molar excess of H_2O_2 in the absence of BH_4 (76). To date, this finding of Dix et al. (75) has not been replicated.

Just as is the case with PAH, our understanding of the mechanism of action of TH will remain incomplete until the role for the protein-bound iron in the enzyme is better understood. The first indication that Fe^{2+} might modulate the activity of the enzyme was the observation that in the presence of tetrahydrofolate, Fe^{2+} could stimulate the conversion of tyrosine to dopa (12). The fact that Fe^{2+} stimulated only slightly when $DMPH_4$ was used in place of tetrahydrofolate was not consistent with a simple interpretation that the hydroxylase is an iron protein.

Subsequently, based on the observation that Fe^{2+} markedly and specifically stimulated the adrenal hydroxylase even in the presence of $DMPH_4$, Petrack et al. (79) concluded that Fe^{2+} "participates" in the hydroxylation reaction.

Studies on the mechanism of the stimulation by added Fe^{2+} of the hydroxylation reaction provided an alternative explanation for the observation of Petrack et al. (79) and indicated that the conclusion that TH is an iron protein was premature. Rather, it was shown that the stimulation by Fe^{2+} was due to its ability to decompose H_2O_2 (322) that is generated by the nonenzymatic oxidation of tetrahydropterins (16). In support of this conclusion, it was shown that TH is inactivated by H_2O_2. Moreover, it was also demonstrated that the stimulation by Fe^{2+} can be mimicked by catalase (16).

It should be emphasized that Shiman et al. (16) did not conclude from their results that TH is *not* an iron protein, but rather that the earlier conclusion that the enzyme was an iron protein (79) was premature and could not be sustained by the data that were then available. Indeed, in the same report that showed that most, if not all, of the observed stimulation of TH by added Fe^{2+} was due to the ability of Fe^{2+} to decompose H_2O_2, it was also demonstrated by titration with o-phenanthroline that partially purified bovine adrenal TH (~40% pure) contains approximately 1.0 mol of iron per subunit (16). Since o-phenanthroline is relatively specific for Fe^{2+}, these results also provided evidence that the protein-bound iron in the resting hydroxylase is in the ferrous state. In this connection, there is evidence that the iron in isolated rat liver PAH, by contrast, is believed to be in the ferric state and that the iron must be reduced to the ferrous state to activate the enzyme (323, 324).

In 1977, Hoeldtke and Kaufman (81), using atomic absorption spectroscopy, showed for the first time that essentially pure TH from bovine adrenal medulla is an iron enzyme and that iron is necessary for catalysis. The highly purified enzyme was found to contain 0.50–0.75 mol of iron per mol of subunit, with the low values probably reflecting some loss of iron from the enzyme during purification. It seems likely that the role of the protein-bound iron in TH will prove to be the same as that of iron in PAH. [See (45) for a discussion of the current status of iron in PAH.]

The observation that TH is an iron protein has often been confirmed. In one of these later studies, "direct" evidence for the presence of iron in rat PC12 TH was obtained with atomic absorption spectroscopy (75). It is not obvious that this evidence was any more "direct" than that reported originally (81), which also employed atomic absorption spectroscopy to measure and quantitate the protein-bound iron. Dix et al. (75), using the Fe^{2+}-specific chelator o-phenanthroline, also confirmed the finding that most of the iron in the resting enzyme is in the Fe^{2+} state, with a stoichiometry of about 1.0 mol of Fe^{2+} per mol of TH subunit being found; no evidence for the presence of Fe^{3+} was obtained with catechol, a chelator that is relatively specific for Fe^{3+}.

These findings, when taken together with the observation that o-phenanthroline and other chelators that are relatively specific for Fe^{2+} are potent inhibitors of the enzyme (57% inhibition at 10 μM o-phenanthroline) (325), whereas chelators that are relatively specific for Fe^{3+} either do not inhibit [e.g., 4,5-hydroxy-1,3-benzenedisulfonic acid (Tiron) (no inhibition at 4 mM) (201)] or inhibit weakly [e.g., catechol (50% inhibition at 5 mM)* (100)] make it extremely likely that enzyme-bound Fe^{2+} is essential for TH activity. The high probability that TH and PAH share a common mechanism also provides indirect support for this conclusion, since the evidence indicates that the enzyme-bound Fe^{2+} participates in the actual hydroxylation step catalyzed by PAH (45).

Despite the strong support for the conclusion that it is the Fe^{2+} form of TH that is catalytically competent, Haavik et al. (201) reported evidence that complicates the picture. They confirmed the presence of iron in their preparation of the pure bovine adrenal enzyme (0.66 ± 0.09 mol iron per mol of subunit) and have also found small amounts of zinc (0.13 mol/mol of subunit). But in addition to these unexceptional results, they found that the enzyme contains an unusual blue-green chromophore (λ_{max} at around 700 nm, $\varepsilon_{270} = 1.9$ mM^{-1} cm^{-1} compared to an ε_{280} of about 60 mM^{-1} cm^{-1}). Results of measurements of the EPR spectra of the enzyme indicate that the chromophore is due to the presence of high-spin Fe^{3+} (in an environment of nearly axial symmetry) with a ligand–iron charge-transfer transition resulting from coordination of catecholate to the iron. Both the intensity of the blue-green color and hydroxylase activity were found to correlate closely with the intensity of the axial type EPR signal.

The conclusion about the nature of the ligands bound to the Fe^{3+} was reinforced and expanded by results of studies of the resonance Raman spectra of the enzyme that indicated the presence of a bidentate Fe^{3+}–catecholate complex. From a comparison of the spectra of TH with that of synthetic iron–catechol and catecholamine complexes, these workers concluded the Fe^{3+} in TH is complexed with one or two carboxylates (anionic ligands) and two or three neutral ligands, such as histidines. Providing some support for the presence of bound catecholamines, it was found that denaturation of the enzyme released significant, but substoichiometric, amounts of norepinephrine (0.11 ± 0.04 mol/mol of subunit) and epinephrine ($0.25 \pm$

---

* The TH activity in brain particles was reported to be more sensitive to inhibition by catechol (41% inhibition at 0.1 mM), but in view of the ability of the pterin coenzyme to reverse the inhibition and the fact that this inhibition study appears to have been carried out in the absence of added tetrahydropterin (12), this degree of inhibition cannot be compared to other values.

0.06 mol/mol of subunit). Since the preparation of TH contained a total of about 0.66 mol of iron per mol of subunit, these results indicate that about one-half of the iron might be complexed with these catecholamines (192). It should be noted that the electron paramagnetic resonance (EPR) spectra of the catecholamine complexes of TH resembles that of PAH (201, 326), an indication that the two enzymes, not surprisingly, have similar iron-coordination sites.

Given the strong evidence that for both TH and PAH it is the protein-bound Fe^{2+} that is involved in catalysis, how does the presence of Fe^{3+} fit into the picture? If the Fe^{3+} represented only a small fraction of the total protein-bound iron, this question would not be a compelling one. What makes it compelling is the conclusion that the EPR-detectable Fe^{3+} represents a "substantial" fraction of the total iron (201). And if a substantial fraction of the iron in TH is present as Fe^{3+}, what role, if any, does it play in catalysis?

First, it is difficult to know how much weight to give to this estimate of the amount of Fe^{3+} in the enzyme since it is based solely on EPR data, an uncertain basis for quantitating the metal content of a protein. In this case, the difficulty may have been compounded by the curious selection of catalase, a heme protein, as the reference standard.

The most convincing evidence against the notion that the Fe^{3+} in TH plays a significant role in catalysis is the finding that it cannot be reduced to Fe^{2+} by a large excess of $6MPH_4$ either aerobically or anaerobically, in the presence or absence of tyrosine, although it can be reduced by millimolar concentrations of dithionite (201). In sharp contrast, the inactive Fe^{3+} form of PAH can be rapidly reduced to Fe^{2+} by stoichiometric amounts of $6MPH_4$ or BH_4 (323, 324). The inability to reduce the Fe^{3+} form of TH under turnover conditions indicates that this "substantial" fraction of the highly active isolated enzyme is catalytically inert.

This conclusion is not entirely mitigated by the observation that phosphorylation of Ser^{40} on TH by PKA, which activates the enzyme, facilitates the reduction of the Fe^{3+} by $6MPH_4$ under turnover conditions (193). Although initial rates of activation and reduction were not measured in this study, the reduction appears to be slow relative to turnover of the enzyme with a half-time of reduction of about 1 min and complete reduction taking as long as 5 min. In the absence of initial rate data, the possibility exists that it is phosphorylation, rather than reduction, that activates the enzyme and that reduction of the protein-bound Fe^{3+} does not contribute significantly to the observed activation. Nonetheless, these results do suggest that the catalytically inactive Fe^{3+} form of the enzyme can be recruited to the pool of active enzyme in a relatively slow, phosphorylation-facilitated reduction.

Based on their confirmation of the earlier observations that Fe^{2+}-specific chelators are potent inhibitors of TH, whereas Fe^{3+}-specific ones (such as Tiron) are not, Haavik et al. (201) at first appeared to endorse the conclusion that the Fe^{2+} form of the enzyme is the functional one. Subsequently, however, citing the ability of certain catechols other than Tiron to inhibit the enzyme (100), these investigators hedged their position and suggested that the Fe^{3+} form of TH does take part in the catalytic cycle (77).

There are, however, other explanations for the sensitivity of TH to certain catechols. First, although it has been dismissed as unlikely (192), chelators of this type, while relatively specific for Fe^{3+}, can also bind Fe^{2+} with low-dissociation constants. For catechol, for example, the dissociation constant for the 1:1 complex with Fe^{2+} is about 0.01 μM (327). If the stability of the complex between catechol and the Fe^{2+} form of TH is similar, it can be calculated that at 10 μM the concentration of catechol that has been reported to inhibit bovine adrenal TH by about 50% at pH 7.0 (192), essentially all of a catalytic amount of the enzyme would be complexed with catechol. Therefore, with a preparation of the enzyme that has one-half of its iron—presumably the Fe^{3+}—complexed with catecholamines (192), exogenous catechol, by combining with the remaining one-half of the iron that is present as Fe^{2+}, could cause significant inhibition.

Catechols could also inhibit TH by combining exclusively with the Fe^{3+} form of the enzyme even if it is the protein-bound Fe^{2+} that takes part in catalysis. It is possible that the combination of catechol with the protein-bound Fe^{3+} induces a conformational change in the enzyme that is unfavorable for its catalytic activity. This possibility can explain the inhibition without the need to implicate the Fe^{3+} form of the enzyme in the catalytic cycle.

Finally, in analogy with what appears to happen during turnover of PAH, where it has been shown that the active Fe^{2+} form of the enzyme reverts back to the inactive Fe^{3+} form (which can be trapped by catechol) once in every 200 turnovers (328), it is likely that some oxidation of Fe^{2+} to Fe^{3+} also occurs during TH turnover. The sensitivity to inhibition by catechol by this mechanism would depend on the length of incubation during the hydroxylase assay and on the rate of reversion per turnover. This mechanism would predict that inhibition by catechols would be time- and turnover dependent.

From the preceding discussion, it is apparent that inhibition by catechol does not necessarily indicate, as has been suggested (77), that the Fe^{3+} form of TH takes part in catalysis. Moreover, there is no evidence that is not compatible with the conclusion that iron plays the same role in catalysis by TH as it does with PAH. [For an outline of the catalytic cycle that has been proposed for PAH, see (45).]

References

1. Kaufman, S., *J. Biol. Chem.*, **230**, 931–939 (1958).
2. Kaufman, S. and Levenberg, B., *J. Biol. Chem.*, **234**, 2683–2688 (1959).
3. Kaufman, S., *Proc. Natl. Acad. Sci. USA*, **50**, 1085–1093 (1963).
4. Blaschko, H. J., *J. Physiol.*, **96**, 50–51 (1939).
5. Goodall, M. and Kirshner, N., *J. Biol. Chem.*, **226**, 213–221 (1957).
6. Gurin, S. and Delluva, A. M., *J. Biol. Chem.*, **170**, 545–550 (1947).
7. Udenfriend, S., Cooper, J. R., Clark, C. T., and Baer, J. E., *Science*, **117**, 663–665 (1953).
8. Udenfriend, S. and Wyngaarden, J. B., *Biochim. Biophys. Acta*, **20**, 48–52 (1956).
9. Levin, E. Y., Levenberg, B., and Kaufman, S., *J. Biol. Chem.*, **235**, 2080–2086 (1960).
10. Brenneman, A. R. and Kaufman, S., *Biochem. Biophys. Res. Commun.*, **17**, 177–183 (1964).
11. Kaufman, S., *Trans. N.Y. Acad. Sci.*, **26**, 977–983 (1964).
12. Nagatsu, T., Levitt, M., and Udenfriend, S., *J. Biol. Chem.*, **239**, 2910–2917 (1964).
13. Perlman, R. L., *Trends Biochem. Sci.*, **4**, N108. (1979).
14. Kaufman, S., The phenylalanine hydroxylating system from mammalian liver, in *Advances in Enzymology*, Vol. 35, Meister, A., Ed., Wiley, New York, 1971, pp. 245–320.
15. Kaufman, S. and Fisher, D. B., Pterin-requiring aromatic amino acid hydroxylases, in *Molecular Mechanisms of Oxygen Activation*, Hayaishi, O., Ed., Academic, New York, 1974, pp. 285–369.
16. Shiman, R., Akino, M., and Kaufman, S., *J. Biol. Chem.*, **246**, 1330–1340 (1971).

17. Kaufman, S., Studies on the mechanism of phenylalanine hydroxylase: detection of an intermediate., in *Chemistry and Biology of Pteridines.*, Pfleiderer, W., Ed., Walter de Gruyter, Berlin, 1975, pp. 291–304.

18. Lazarus, R. A., Benkovic, S. J., and Kaufman, S., *J. Biol. Chem.*, **258**, 10960–10962 (1983).

19. Udenfriend, S., *Harvey Lecture Ser.*, **60**, 57–83 (1964).

20. Laduron, P. and Belpaire, F., *Nature (London)*, **217**, 1155–1156 (1968).

21. Wurzburger, R. J. and Musacchio, J. M., *J. Pharmacol. Exp. Ther.*, **177**, 155–168 (1971).

22. Vigny, A., Flamand, M.-F., and Henry, J.-P., *FEBS Lett.*, **86**, 235–238 (1978).

23. McGeer, E. G., Gibson, S., and McGeer, P. L., *Can. J. Biochem.*, **45**, 1557–1563 (1967).

24. McGeer, E. G., McGeer, P. L., and Wada, J. A., *J. Neurochem.*, **18**, 1647–1658 (1971).

25. Kuczenski, R. T. and Mandell, A. J., *J. Biol. Chem.*, **247**, 3114–3122 (1972).

26. Segal, D. S. and Kuczenski, R., *Brain Res.*, **68**, 261–266 (1974).

27. Pickel, V. M., Joh, T. H., and Reis, D. J., *Proc. Natl. Acad. Sci. USA*, **72**, 659–663 (1975).

28. Coyle, J. T., *Biochem. Pharmacol.*, **21**, 1935–1944 (1972).

29. Bullard, W. P., Guthrie, P. B., Russo, P. V., and Mandell, A. J., *J. Pharmacol. Exp. Thera.*, **206**, 4–20 (1978).

30. Levine, R. A., Kuhn, D. M., and Lovenberg, W., *J. Neurochem.*, **32**, 1575–1578 (1979).

31. Bredt, D. S. and Snyder, S. H., *Proc. Natl. Acad. Sci. USA*, **87**, 682–685 (1990).

32. Giovanelli, J., Campos, K. L., and Kaufman, S., *Proc. Natl. Acad. Sci. USA*, **88**, 7091–7095 (1991).

33. Craine, J. E., Hall, E. S., and Kaufman, S., *J. Biol. Chem.*, **247**, 6082–6091 (1972).

34. Pomerantz, S. H., *Biochem. Biophys. Res. Commun.*, **16**, 188–194 (1964).

35. Karobath, M., *Proc. Natl. Acad. Sci. USA*, **68**, 2370–2373 (1971).

36. Reinhard, J. F., Smith, G. K., and Nichol, C. A., *Life Sci.*, **39**, 2185–2189 (1986).

37. Waymire, J. C., Bjur, R., and Weiner, N., *Anal. Biochem.*, **43**, 588–600 (1971).

38. Ichiyama, A., Nakamura, S., Mishizuka, Y., and Hayaishi, O., *J. Biol. Chem.*, **245**, 1699–1709 (1970).

39. Keller, R., Oke, A., Mefford, I., and Adams, R. N., *Life Sci.*, **19**, 995–1003 (1976).

40. Felice, L. J., Felice, J. D., and Kissinger, P. T., *J. Neurochem.*, **31**, 1461–1465 (1978).

41. Sasa, S. and Blank, C. L., *Anal. Chem.*, **49**, 354–359 (1977).

42. Nielsen, J. A. and Johnston, C. A., *Life Sci.*, **31**, 2847–2856 (1982).

43. Messripour, M. and Clark, J. B., *J. Neurochem.*, **38**, 1139–1143 (1982).

44. Kato, T., Horiuchi, S., Togari, A., and Nagatsu, T., *Experientia, 37*, 809–811 (1981).

45. Kaufman, S., The phenylalanine hydroxylating system, in *Advances in Enzymology*, Vol. 67, Meister, A., Ed., Wiley, New York, 1993, pp. 77–264.

46. Lloyd, T., Mori, T., and Kaufman, S., *Biochemistry, 10*, 2330–2336 (1971).

47. Ellenbogen, L., Taylor, R. J., and Brundage, G. B., *Biochem. Biophys. Res. Commun., 19*, 708–715 (1965).

48. Shiman, R. and Kaufman, S., Tyrosine hydroxylase, in *Methods in Enzymology*, Vol. 17A, Tabor, H. and Tabor, C. W., Eds., Academic, New York, 1970, pp. 609–615.

49. Numata (Sudo), Y., Kato, T., Nagatsu, T., Sugimoto, T., and Matsuura, S., *Biochim. Biophys. Acta, 480*, 104–112 (1977).

50. Kato, T., Oka, K., Nagatsu, T., Sugimoto, T., and Matsuura, S., *Biochim. Biophys. Acta, 612*, 226–232 (1980).

51. Pollock, R. J., Kapatos, G., and Kaufman, S., *J. Neurochem., 37*, 855–860 (1981).

52. Lazar, M. A., Lockfeld, A. J., Truscott, R. J. W., and Barchas, J. D., *J. Neurochem., 39*, 409–422 (1982).

53. Kuczenski, R., *J. Biol. Chem., 248*, 5074–5080 (1973).

54. Okuno, S. and Nakata, H., *Eur. J. Biochem., 122*, 49–55 (1982).

55. Markey, K. A., Kondo, S., Shenkman, L., and Goldstein, M., *Mol. Pharmacol., 17*, 79–85 (1980).

56. Bailey, S. W., Dillard, S. B., Thomas, K. B., and Ayling, J. E., *Biochemistry, 28*, 494–504 (1989).

57. Lloyd, T. and Kaufman, S., *Biochem. Biophys. Res. Commun., 66*, 907–917 (1975).

58. Miller, L. P. and Lovenberg, W., *Neurochem. Int., 7*, 689–697 (1985).

59. Kaufman, S., *Biochim. Biophys. Acta, 51*, 619–621 (1961).

60. Storm, C. B. and Kaufman, S., *Biochem. Biophys. Res. Commun., 32*, 788–793 (1968).

61. Ikeda, M., Levitt, M., and Udenfriend, S., *Biochem. Biophys. Res. Commun., 18*, 482–488 (1965).

62. Tong, J. H., D'Iorio, A., and Benoiton, N. L., *Biochem. Biophys. Res. Commun., 44*, 229–236 (1971).

63. Tong, J. H., D'Iorio, A., and Benoiton, N. L., *Biochem. Biophys. Res. Commun., 43*, 819–826 (1971).

64. Kaufman, S., Regulatory properties of tyrosine hydroxylase, in *Neurobiological Mechanisms of Adaptation and Behavior* Vol. 13, Mandell, A. J., Ed., Raven Press, New York, 1975, pp. 127–136.

65. Oka, K., Kato, T., Sugimoto, T., Matsuura, S., and Nagatsu, T., *Biochim. Biophys. Acta, 661*, 45–53 (1981).

66. Lloyd, T. and Kaufman, S., *Biochem. Biophys. Res. Commun.*, **59**, 1262–1270 (1974).

67. Katz, I. R., Yamauchi, T., and Kaufman, S., *Biochim. Biophys. Acta,* **429**, 84–95 (1976).

68. Bagchi, S. P. and Zarycki, E. P., *Life Sci.*, **9**, 111–119 (1970).

69. Bagchi, S. P. and Zarycki, E. P., *Biochem. Pharmacol.*, **22**, 1353–1368 (1973).

70. Karobath, M. and Baldessarini, R. J., *Nat. New Biol.*, **236**, 206–208 (1972).

71. Katz, I., Lloyd, T., and Kaufman, S., *Biochim. Biophys. Acta,* **445**, 567–578 (1976).

72. Haavik, J. and Flatmark, T., *Eur. J. Biochem.*, **168**, 21–26 (1987).

73. Fitzpatrick, P. F., *Biochemistry*, **30**, 6386–6391 (1991).

74. Fukami, M. H., Haavik, J., and Flatmark, T., *Biochem. J.*, **268**, 525–528 (1990).

75. Dix, T. A., Kuhn, D. M., and Benkovic, S. J., *Biochemistry*, **26**, 3354–3361 (1987).

76. Ribeiro, P., Pigeon, D., and Kaufman, S., *J. Biol. Chem.*, **266**, 16207–16211 (1991).

77. Andersson, K. K., Vassort, C., Brennan, B. A., Que, L., Jr., Haavik, J., Flatmark, T., Gros, F., and Thibault, J., *Biochem. J.*, **284**, 687–695 (1992).

78. Kaufman, S., The enzymology of phenylketonuria, in *ILSI-Nutrition Foundation Aspartame Workshop Proceedings, Marbella, Spain, Nov. 17–21, 1986,* International Life Sciences Institute, 1986, pp. 2–39.

79. Petrack, B., Sheppy, F., and Fetzer, V., *J. Biol. Chem.*, **243**, 743–748 (1968).

80. Musacchio, J. M., Wurtzburger, R. J., and D'Angelo, G. L., *Mol. Pharmacol.*, **7**, 136–146 (1971).

81. Hoeldtke, R. and Kaufman, S., *J. Biol. Chem.*, **252**, 3160–3169 (1977).

82. Fisher, D. B. and Kaufman, S., *J. Biol. Chem.*, **248**, 4345–4353 (1973).

83. Joh, T. H. and Reis, D. J., *Brain Res.*, **85**, 146–151 (1975).

84. Grima, B., Lamouroux, A., Blanot, F., Biguet, N. F., and Mallet, J., *Proc. Natl. Acad. Sci. USA,* **82**, 617–621 (1985).

85. Kaufman, S. and Kaufman, E. E., Tyrosine hydroxylase, in *Folates and Pterins, Chemistry and Biochemistry of Pterins,* Vol. 2, Blakely, R. L. and Benkovic, S. J., Eds., Wiley, New York. 1985 pp. 251–352.

86. Musacchio, J. M., McQueen, C. A., and Craviso, G. L., Tyrosine hydroxylase: Subcellular distribution and molecular and kinetic characteristics of the different enzyme forms, in *New Concepts in Neurotransmitter Regulation,* Mandell, A. J., Ed., Plenum, New York, 1973, pp. 69–88.

87. Lloyd, T., *J. Biol. Chem.*, **254**, 7247–7254 (1979).

88. Raese, J., Patrick, R. L., and Barchas, J. D., *Biochem. Pharmacol.*, **25**, 2245–2250 (1976).

89. Vigny, A. and Henry, J.-P., *J. Neurochem.*, **36**, 483–489 (1981).

90. Kaufman, S., Properties of the pterin-dependent aromatic amino acid hydroxyl-

ases, in *Aromatic Amino Acids in the Brain, Ciba Foundation Symposium 22,* Wolstenholme, G. E. W. and Fitzsimmons, D. W., Eds., Elsevier Excerpta Medica, North-Holland, Amsterdam, 1974, pp. 85–108.

91. Kuczenski, R., Rat brain tyrosine hydroxylase. *J. Biol. Chem.,* **248,** 2261–2265 (1973).

92. Holland, W. C. and Schuman, H. J., *Br. J. Pharmacol. Chemother.,* **4,** 449–453 (1956).

93. Butterworth, K. R. and Mann, M., *Br. J. Pharmacol. Chemother.,* **12,** 426 (1957).

94. Bygdeman, S. and von Euler, U. S., *Acta Physiol. Scand.,* **44,** 375–383 (1958).

95. Roth, R. H., Stjarne, L., and von Euler, U. S., *Life Sci.,* **5,** 1071–1075 (1966).

96. Alousi, A. and Weiner, N., *Proc. Natl. Acad. Sci. USA,* **56,** 1491–1496 (1966).

97. Gordon, R., Reid, J. V. D., Sjoerdsma, A., and Udenfriend, S., *Mol. Pharmacol.,* **2,** 610–613 (1966).

98. Roth, R. H., Stjarne, L., and von Euler, U. S., *J. Pharmacol. Exp. Ther.,* **158,** 373–377 (1967).

99. Sedvall, G. C. and Kopin, I. G., *Life Sci.,* **6,** 45–51 (1967).

100. Udenfriend, S., Zaltzman-Nirenberg, P., and Nagatsu, J., *Biochem. Pharmacol.,* **14,** 837–845 (1965).

101. Spector, S., Gordon, R., Sjoerdsma, A., and Udenfriend, S., *Mol. Pharmacol.,* **3,** 549–555 (1967).

102. Weiner, N. and Rabadjija, M., *J. Pharmacol. Exp. Ther.,* **160,** 61–71 (1968).

103. Cloutier, G. and Weiner, N., *J. Pharmacol. Exp. Ther.,* **186,** 75–85 (1973).

104. Zivkovic, G., Guidotti, A., and Costa, E., *Mol. Pharmacol.,* **10,** 727–735 (1974).

105. Reis, D. J., Joh, T. H., Ross, R. A., and Pickel, V. M., *Brain Res.,* **81,** 380–386 (1974).

106. Morgenroth, V. H. I., Boadle-Biber, M., and Roth, R. H., *Proc. Natl. Acad. Sci. USA,* **71,** 4283–4287 (1974).

107. Kakiuchi, S., Rall, T., and Mcllwain, H., *J. Neurochem.,* **16,** 485–491 (1969).

108. Schimizu, H., Creveling, C. R., and Daly, J. W., *Mol. Pharmacol.,* **6,** 184–188 (1970).

109. Goldstein, M., Anagnoste, B., and Shirron, C., *J. Pharm. Pharmacol.,* **25,** 348–351 (1973).

110. Harris, J. E., Morgenroth, V. H., Roth, R. H., and Baldessarini, R. J., *Nature (London),* **252,** 156–158 (1974).

111. Morgenroth, V. H. I., Hegstrand, L. R., Roth, R. H., and Greengard, P., *J. Biol. Chem.,* **250,** 1946–1948 (1975).

112. Lovenberg, W., Bruckwick, E., and Hanbauer, I., Protein phosphorylation and regulation of catecholamine synthesis, in *Chemical Tools in Catecholamine Research,* Vol. 2, Almgren, O. and Carlsson, A., Eds., North-Holland Publishing Co., Amsterdam, 1975, pp. 37–44.

113. Lovenberg, W., Bruckwick, E. A., and Hanbauer, I., *Proc. Natl. Acad. Sci. USA*, **72**, 2955–2958 (1975).

114. Yamauchi, T. and Fujisawa, H., *J. Biol. Chem.*, **254**, 503–507 (1979).

115. Yamauchi, T. and Fujisawa, H., *J. Biol. Chem.*, **254**, 6408–6413 (1979).

116. Vulliet, P. R., Langan, T. A., and Weiner, N., *Proc. Natl. Acad. Sci. USA*, **77**, 92–96 (1980).

117. Vrana, K. E., Allhiser, C. L., and Roskoski, J. R., *J. Neurochem.*, **30**, 92–100 (1981).

118. Masserano, J. M. and Weiner, N., *Mol. Pharmacol.*, **16**, 513–528 (1979).

119. Vrana, K. E. and Roskoski, J., R., *J. Neurochem.*, **40**, 1692–1700 (1983).

120. Bullard, W. P. and Capson, T. L., *Mol. Pharmacol.*, **23**, 104–111 (1983).

121. Joh, T. H., Park, D. H., and Reis, D. J., *Proc. Natl. Acad. Sci. USA*, **75**, 4744–4748 (1978).

122. Edelman, A. M., Raese, J. D., Lazar, M. A., and Barchas, J. D., *J. Pharmacol. Exp. Ther.*, **216**, 647–653 (1981).

123. Richtand, N. M., Inagami, T., Misono, K., and Kuczenski, R., *J. Biol. Chem.*, **260**, 8465–8473 (1985).

124. Nelson, T. J. and Kaufman, S., *Arch. Biochem. Biophys.*, **257**, 69–84 (1987).

125. Ames, M. M., Lerner, P., and Lovenberg, W., *Biochem. J.*, **253**, 27–31 (1978).

126. Hegstrand, L. R., Simon, J. R., and Roth, R. H., *Biochem. Pharmacol.*, **28**, 519–523 (1979).

127. Goldstein, M., Bronaugh, R. L., Ebstein, B., and Rogerge, C., *Brain Res.*, **109**, 563–574 (1976).

128. Weiner, N., *Monogr. Neural Sci.*, **7**, 146–160 (1980).

129. Murrin, L. C., Morgenroth, V. H., and Roth, R. H., *Mol. Pharmacol.*, **12**, 1070–1081 (1976).

130. Weiner, N., Lee, F., Dreyer, E., and Barnes, E., *Life Sci.*, **22**, 1197–1216 (1978).

131. Abita, J. P., Milstien, S., Chang, N., and Kaufman, S., *J. Biol. Chem.*, **251**, 5310–5314 (1976).

132. Milstien, S., Abita, J. P., Chang, N., and Kaufman, S., *Proc. Natl. Acad. Sci. USA*, **73**, 1591–1593 (1976).

133. Krueger, B. K., Forn, J., and Greengard, P., *J. Biol. Chem.*, **252**, 2764–2773 (1977).

134. Nathanson, J. A., Cyclic nucleotides and nervous system function, in *Physiological Reviews*, Vol. 57, Morgan, H. E. and Rose, R. C., Eds., American Physiological Society, Bethesda, MD, 1977, pp. 158–213.

135. Bustos, G. and Roth, R. H., *Biochem. Pharmacol.*, **28**, 3026–3028 (1979).

136. Baker, P. F., Hodgkin, A. L., and Ridgway, E. B., *J. Physiol.*, **218**, 709–755 (1971).

137. Patrick, R. B. and Barchas, J. D., *Nature (London)*, **250**, 737–739 (1974).

138. El Mestikawy, S., Glowinski, J., and Hamon, M., *Nature (London)*, **302**, 830–832 (1983).

139. Nose, P. S., Griffith, L. C., and Schulman, H., *J. Cell Biol.*, **101**, 1182–1190 (1985).

140. Yamauchi, T. and Fujisawa, H., *Biochem. Int.*, **1**, 98–104 (1980).

141. Yamauchi, T. and Fujisawa, H., *Biochem. Biophys. Res. Commun.*, **100**, 807–813 (1981).

142. Atkinson, J., Richtand, N., Schworer, C., Kuczenski, R., and Soderling, T., *J. Neurochem.*, **49**, 1241–1249 (1987).

143. Yamauchi, T., Nakata, H., and Fujisawa, H., *J. Biol. Chem.*, **256**, 5404–5409 (1981).

144. Yamauchi, T. and Fujisawa, H., *FEBS Lett.*, **129**, 117–119 (1981).

145. Moore, B. W. and Perez, V. J., Specific acidic proteins of the nervous system, in *Physiological and Biochemical Aspects of Nervous Integration,* Carlson, F. D., Ed., Prentice-Hall, Englewood Cliff, NJ, 1967, pp. 343–359.

146. Ichimura, T., Isobe, T., Okuyama, T., Yamauchi, T., and Fujisawa, H., *FEBS Lett.*, **219**, 79–82 (1987).

147. Aitken, A., Ellis, C. A., Harris, A., Sellers, L. A., and Toker, A., *Nature (London)*, **344**, 594 (1990).

148. Morgenroth, H. I., Boadle-Biber, M. C., and Roth, R. H., *Trans. Am. Soc. Neurochem.*, **5**, 78 (1974).

149. Lerner, P., Ames, M. M., and Lovenberg, W., *Mol. Pharmacol.*, **13**, 44–49 (1977).

150. Kapatos, G. and Zigmond, M. J., *Brain Res.*, **170**, 299–312 (1979).

151. Fujisawa, H., Yamauchi, T., Nakata, H., and Okuno, S., *Fed. Proc.*, **43**, 3011–3014 (1984).

152. McGuinness, T. L., Lai, Y., Greengard, P., Woodgett, J. P., and Cohen, P., *FEBS Lett.*, **163**, 329–334 (1983).

153. Payne, M. E., Schworer, C. M., and Soderling, T. R., *J. Biol. Chem.*, **258**, 2376–2382 (1983).

154. Ahmad, Z., DePaoli-Roach, A. A., and Roach, P. J., *J. Biol. Chem.*, **257**, 8348–8355 (1982).

155. Vulliet, P. R., Woodgett, J. R., and Cohen, P., *J. Biol. Chem.*, **259**, 13680–13683 (1984).

156. Takai, Y., Kishimoto, A., Iwasa, Y., Kawahara, Y., Mori, T., and Nishizuka, Y., *J. Biol. Chem.*, **254**, 3692–3695 (1979).

157. Raese, J. D., Edelman, A. M., Makk, G., Bruckwick, E. A., Lovenberg, W., and Barchas, J. D., *Commun. Psychopharmacol.*, **3**, 295–301 (1979).

158. Albert, K. A., Helmer-Matyjek, E., Nairn, A. C., Müller, T. H., Haycock, J. W., Greene, L. A., Goldstein, M., and Greengard, P., *Proc. Natl. Acad. Sci. USA*, **81**, 7713–7717 (1984).

159. Andrews, D. W., Langan, T. A., and Weiner, N., *Proc. Natl. Acad. Sci. USA*, **80**, 2097–2101 (1983).

160. Roskoski, R., Jr., Vulliet, P. R., and Glass, D. B., *J. Neurochem.*, **48**, 840–845 (1987).

161. Haycock, J. W., Bennett, W. F., George, R. J., and Waymire, J. C., *J. Biol. Chem.*, **257**, 13699–13703 (1982).

162. Amy, C. M. and Kirshner, N. J., *J. Neurochem.*, **36**, 847–854 (1981).

163. Haycock, J. W., Meligeni, J. A., Bennett, W. F., and Waymire, J. C., *J. Biol. Chem.*, **257**, 12641–12648 (1982).

164. Kuo, J. F., Andersson, R. G. G., Wise, R. C., Mackerlova, L., Salomonsson, I., Brackett, N. L., Shoji, M., and Wrenn, R. W., *Proc. Natl. Acad. Sci. USA*, **77**, 7039–7043 (1980).

165. Kurosawa, A., Guidotti, A., and Costa, E., *Mol. Pharmacol.*, **12**, 420–432 (1976).

166. Levi-Montalcini, R. and Angelletti, P. U., *Physiol. Rev.*, **48**, 534–569 (1968).

167. Halegoua, S. and Patrick, J., *Cell*, **22**, 571–581 (1980).

168. McTigue, M., Cremins, J., and Halegoua, S., *J. Biol. Chem.*, **260**, 9047–9056 (1985).

169. Yamanishi, J., Kaibuchi, K., Sano, K., Castagna, M., and Nishizuka, Y., *Biochem. Biophys. Res. Commun.*, **112**, 778–786 (1983).

170. Lee, K. Y., Seeley, P. J., Muller, T. H., Helmer-Matyjek, E., Sabban, E., Goldstein, M., and Greene, L. A., *Mol. Pharmacol.*, **28**, 220–228 (1985).

171. Cremins, J., Wagner, J. A., and Halegoua, S., *J. Cell Biol.*, **103**, 887–893 (1986).

172. Griffith, L. C. and Schulman, H., *J. Biol. Chem.*, **19**, 9542–9549 (1988).

173. Yanagihara, N., Tank, A. W., Langan, T. A., and Weiner, N., *J. Neurochem.*, **46**, 562–568 (1986).

174. Waymire, J. C., Johnston, J. P., Hummer-Lickteig, K., Lloyd, A., Vigny, A., and Craviso, G. L., *J. Biol. Chem.*, **263**, 12439–12447 (1988).

175. Cahill, A. L. and Perlman, R. L., *Biochim. Biophys. Acta*, **805**, 217–226 (1984).

176. Vulliet, P. R., Woodgett, J. R., Ferrari, S., and Hardie, D. G., *FEBS Lett.*, **182**, 335–339 (1985).

177. Campbell, D. G., Hardie, D. G., and Vulliet, P. R., *J. Biol. Chem.*, **261**, 10489–10492 (1986).

178. Vulliet, P. R., Hall, F. L., Mitchell, J. P., and Hardie, D. G., *J. Biol. Chem.*, **264**, 16292–16298 (1989).

179. Wretborn, M., Humble, E., Ragnarsson, U., and Engström, L., *Biochem. Biophys. Res. Commun.*, **93**, 403–408 (1980).

180. Dahl, H.-H. M., and Mercer, J. F. B., *J. Biol. Chem.*, **261**, 4148–4153 (1986).

181. Haycock, J. W., *J. Biol. Chem.*, **265**, 11682–11691 (1990).

182. Haycock, J. W., and Haycock, D. A., *J. Biol. Chem.*, **266**, 5650–5657 (1991).

183. Mitchell, J. P., Hardie, D. G., and Vulliet, P. R., *J. Biol. Chem.*, **265**, 22358–22364 (1990).

184. Haycock, J. W., Ahn, N. G., Cobb, M. H., and Krebs, E. G., *Proc. Natl. Acad. Sci. USA*, **89**, 2365–2369 (1992).

185. Anderson, N. G., Maller, J. L., Tonks, N. K., and Sturgill, T. W., *Nature* (*London*), **343**, 651–653 (1990).

186. Boulton, T. G., Nye, S. H., Robbins, D. J., Ip, N. Y., Radziejewska, E., Morgenbesser, S. D., DePinho, R. A., Panayotatos, N., Cobb, M. H., and Yancopoulos, G. D., *Cell,* **65,** 663–675 (1991).

187. Kapatos, G. and Kaufman, S., Inhibition of pterin biosynthesis in the adrenergic neuroblastoma NIE115 by tetrahydrobiopterin and folate, in *Chemistry and Biology of Pteridines,* Blair, J. A., Ed., Walter de Gruyter, Berlin, 1983, pp. 171–175.

188. Funakoshi, H., Okuno, S., and Fujisawa, H., *J. Biol. Chem.,* **266,** 15614–15620 (1991).

189. Kaufman, S., Hasegawa, H., Wilgus, H., and Parniak, M., Regulation of hepatic phenylalanine hydroxylase activity by phosphorylation and dephosphorylation, in *Cold Spring Harbor Conferences on Cell Proliferation* (*Protein Phosphorylation*), Vol. 8, Laboratory, C. S. H., Ed., Cold Spring Harbor Conferences, Cold Spring Harbor, NY, 1981, pp. 1391–1406.

190. Okuno, S. and Fujisawa, H., *Biochem. Biophys. Res. Commun.,* **124,** 223–228 (1984).

191. Okuno, S. and Fujisawa, H., *J. Biol. Chem.,* **260,** 2633–2635 (1985).

192. Andersson, K. K., Cox, D. D., Que, L., Jr., Flatmark, T., and Haavik, J., *J. Biol. Chem.,* **263,** 18621–18626 (1988).

193. Andersson, K. K., Haavik, J., Martinez, A., Flatmark, T., and Petersson, L., *FEBS Lett.,* **258,** 9–12 (1989).

194. Haavik, J., Martiney, A., and Flatmark, T., *FEBS Lett.,* **262,** 363–365 (1990).

195. Edelman, A. M., Blumenthal, D. K., and Krebs, E. G., Protein serine/threonine kinases, in *Annual Reviews of Biochemistry.,* Vol. 56, Richardson, C. C. Boyer, P. D. Dawid, I. B. and Meister, A., Eds., Annual Reviews, Inc., Palo Alto, CA, 1987, pp. 567–613.

196. Zigmond, R. E., Schwarzschild, M. A., and Rittenhouse, A. R., *Ann. Rev. Neurosci.,* **1989,** 415–461 (1989).

197. LeBourdelles, B., Horellou, P., LeCaer, J.-P., Denefle, P., Latta, M., Haavik, J., Guibert, B., Mayaux, J.-F., and Mallet, J., *J. Biol. Chem.,* **266,** 17124–17130 (1991).

198. Daubner, S. C., Lauriano, C., Haycock, J. W., and Fitzpatrick, P. F., *J. Biol. Chem.,* **267,** 12639–12646 (1992).

199. Wang, Y.-H., Citron, B., Ribeiro, P., and Kaufman, S., *Proc. Natl. Acad. Sci. USA,* **88,** 8779–8783 (1991).

200. Ribeiro, P., Wang, Y.-H., Citron, B. A., and Kaufman, S., *Proc. Natl. Acad. Sci. USA,* **89,** 9593–9597 (1992).

201. Haavik, J., Andersson, K. K., Petersson, L., and Flatmark, T., *Biochim. Biophys. Acta,* **953,** 142–156 (1988).

202. Roth, R. H. and Salzman, P. M., Role of calcium in the depolarization-induced activation of tyrosine hydroxylase, in *Structure and Function of Monoamine*

Enzymes, Usdin, E. Weiner, N. and Youdim, M. B. H., Eds., Marcel-Decker, New York, 1977, pp. 149–168.

203. Thompson, T. L., Colby, K. A., and Patrick, R. L., *Neurochemical Research* **15,** 1159–1166 (1990).

204. Fujisawa, H., Yamauchi, T., Nakata, H., and Okuno, S., Regulation of tryptophan 5-monooxygenase and tyrosine 3-monooxygenase by protein kinases, in *Oxygenases and Oxygen Metabolism,* Nozaki, M. Yamamoto, S. Ishimura, Y. Coon, M. J. Ernster, L. and Estabrook, R. W., Eds., Academic, New York, 1982, pp. 281–292.

205. Erny, R. E., Berezo, M. W., and Perlman, R. L., *J. Biol. Chem.,* **256,** 1335–1339 (1981).

206. Roskoski, R., Jr. and Roskoski, L. M., *J. Neurochem.,* **53,** 1934–1940 (1989).

207. Yanagihara, N., Tank, A. W., and Weiner, N., *Mol. Pharmacol.,* **26,** 141–147 (1984).

208. Simon, J. R. and Roth, R. H., *Mol. Pharmacol.,* **16,** 224–233 (1979).

209. Weiner, N., Lee, F.-L., Barnes, E., and Dreyer, E., Enzymology of tyrosine hydroxylase and the role of cyclic nucleotides in its regulation, in *Structure and Function of Monamine Enzymes,* Usdin, E. Weiner, N. and Youdim, M., Eds., Dekker, New York, 1977, pp. 109–148.

210. Iuvone, P. M., Galli, C. L., and Neff, N. H., *Mol. Pharmacol.,* **14,** 1212–1219 (1978).

211. Roth, R. H., Walters, J. R., and Aghajanian, G. K., Effects of impulse flow on the release and synthesis of dopamine in the rat striatum, in *Frontiers in Catecholamine Research,* Usdin, E. and Snyder, S. H., Eds., Pergamon, New York, 1973, pp. 567–574.

212. Almas, B., LeBourdelles, B., Flatmark, T., Mallet, J., and Haavik, J., *Eur. J. Biochem.,* **209,** 249–255 (1992).

213. Phillips, R. S., and Kaufman, S., *J. Biol. Chem.,* **259,** 2474–2479 (1984).

214. Døskeland, A. P., Døskeland, S. O., Ogreid, D., and Flatmark, T., *J. Biol. Chem.,* **257,** 11242–11248 (1984).

215. Nelson, T. J. and Kaufman, S., *J. Biol. Chem.,* **262,** 16470–16475 (1987).

216. Ribeiro, P. and Kaufman, S., *Neurochem. Res.,* **19,** 541–548 (1994).

217. Haavik, J., Schelling, D. L., Campbell, D. G., Andersson, K. K., Flatmark, T., and Cohen, P., *FEBS Lett.,* **251,** 36–42 (1989).

218. Shiman, R., Mortimore, G. E., Schworer, C. M., and Gray, D. W., *J. Biol. Chem.,* **257,** 11213–11216 (1982).

219. Thoenen, H., *Nature (London),* **228,** 861–862 (1970).

220. Kvetnansky, R., Weise, V. K., Gewirtz, G. P., and Kopin, I. J., *Endocrinology,* **89,** 46–49 (1971).

221. Reis, D. J., Moorhead, D. T., Rifkin, M., Joh, T. H., and Goldstein, M., *Nature (London),* **229,** 562–563 (1971).

222. Mueller, R. A., Thoenen, H., and Axelrod, J., *J. Pharmacol. Exp. Ther.,* **169,** 74–79 (1969).

223. Edgar, D. H. and Thoenen, H., *Brain Res.*, **154**, 186–190 (1978).

224. Lucas, C. A. and Thoenen, H., *Neuroscience*, **2**, 1095–1101 (1977).

225. Acheson, A. L., Naujoks, K., and Thoenen, H., *J. Neurosci.*, **4**, 1771–1780 (1984).

226. Goodman, R. and Herschman, H. R., *Proc. Natl. Acad. Sci. USA*, **75**, 4587–4590 (1978).

227. Naujoks, K. W., Korsching, S., Rohrer, H., and Thoenen, H., *Dev. Biol.*, **92**, 365–379 (1982).

228. Goodman, R., Slater, E., and Herschman, H. R., *J. Cell Biol.*, **84**, 495–500 (1980).

229. Zigmond, R. E. and Ben-Ari, Y., *Proc. Natl. Acad. Sci. USA*, **74**, 3078–3080 (1977).

230. Chalazonitis, A., Rice, P. J., and Zigmond, R. E., *J. Pharmacol. Exp. Ther.*, **213**, 139–143 (1980).

231. Kumakara, K., Guidotti, A., and Costa, E., *Mol. Pharmacol.*, **16**, 865–876 (1979).

232. Mueller, R. A. H., Thoenen, H., and Axelrod, J., *Science*, **158**, 468–469 (1969).

233. Thoenen, H. and Tranzer, J. P., *Naunyn-Schmiedebergs Arch. Pharmakol. Exp. Pathol.*, **261**, 271–288 (1968).

234. Thoenen, H., Mueller, R. A., and Axelrod, J., *J. Pharmacol. Exp. Ther.*, **169**, 249–254 (1969).

235. Joh, T. H., Geghman, C., and Reis, D. J., *Proc. Natl. Acad. Sci. USA*, **70**, 2767–2771 (1973).

236. Hoeldtke, R., Lloyd, T., and Kaufman, S., *Biochem. Biophys. Res. Commun.*, **57**, 1045–1053 (1974).

237. Chuang, D. M. and Costa, E., *Proc. Natl. Acad. Sci. USA*, **71**, 4570–4574 (1974).

238. Zigmond, R. E., Schon, R., and Iversen, L. L., *Brain Res.*, **70**, 547–552 (1974).

239. Thoenen, H. and Otten, V., Trans-synaptic induction of tyrosine hydroxylase in superior cervical ganglia: Participation of postsynaptic on steroid receptors, in *Chemical Tools in Catecholamine Research*, Vol. 2, Almgren, O. and Carlsson, A., Eds., North-Holland Publishing Co., Amsterdam, 1975, pp. 175–182.

240. Guidotti, A. and Costa, E., *J. Pharmacol. Exp. Ther.*, **189**, 665–675 (1974).

241. Costa, E., Chuang, D. M., and Guidotti, A., Adrenal medulla: A model to study the trans-synaptic regulation of gene expression, in *Structure and Function of Monoamine Enzymes*, Usdin, E. Weiner, N. and Youdim, M. B. H., Eds., Marcel-Dekker, New York, 1977, pp. 279–310.

242. Guidotti, A., Chuang, D. M., Hollenbeck, R., and Costa, E., Nuclear translocation of catalytic subunits of cytosol cAMP-dependent protein kinase in the transsynaptic induction of medullary tyrosine hydroxylase, in *Advances in Cyclic Nucleotide Research*, Vol. 9, George, W. J. and Ignarro, L. J., Eds., Raven Press, New York, 1978, pp. 185–197.

243. Costa, E., Chuang, D. M., Guidotti, A., and Uzunov, P., Cyclic 3'.5' adenosine monophosphate dependent molecular mechanisms in the trans-synaptic induction of tyrosine hydroxylase in rat adrenal medulla, in *Chemical Tools in Catecholamine Research,* Vol. 2, Almgren, O. and Carlsson, A., Eds., North-Holland Publishing Co., Amsterdam, 1975, pp. 283–290.

244. Palmer, W. K., Castagna, M., and Walsh, D. A., *Biochem. J.,* **143,** 469–471 (1974).

245. Jungmann, R. A., Hiestand, P. C., and Schweppe, J. S., *Endocrinology.,* **94,** 168–183 (1974).

246. Hanbauer, I. and Costa, B., Trans-synaptic induction of tyrosine hydroxylase in superior cervical ganglia: participation of postsynaptic and steroid receptors, in *Chemical Tools in Catecholamine Research,* Vol. 2, Almgren, O. and Carlsson, A., Eds., North-Holland Publishing Co., Amsterdam, The Netherlands, 1975, pp. 175–182.

247. Waymire, J. C., Weiner, N., and Prasad, K. N., *Proc. Natl. Acad. Sci. USA,* **69,** 2241–2245 (1972).

248. Williams, L. R., Sandquist, D., Black, A. C., and Williams, T. H., *J. Neurochem.,* **36,** 2057–2062 (1981).

249. Tank, A. W. and Weiner, N., *Mol. Pharmacol.,* **22,** 421–430 (1982).

250. Baetge, E. E., Kaplan, B. B., Reis, D. J., and Joh, T. H., *Proc. Natl. Acad. Sci. USA,* **78,** 1269–1273 (1981).

251. Yamamoto, K. R. and Alberts, B. M., *Annu. Rev. Biochem.,* **45,** 721–746 (1976).

252. Rosenfeld, M. G. and Barrieux, A., *Adv. Cyclic Nucl. Res.,* **11,** 205–264 (1979).

253. Wicks, W. D., Leichtling, B. H., Wimalasena, J., Roper, M. D., Su, J. L., Su, Y.-F., Howell, S., Harden, T. K., and Wolfe, B. B., *Adv. Cyclic Nucl. Res.,* **9,** 411–424 (1978).

254. Thoenen, H., Angeletti, P. U., Levi-Montalcini, R., and Kettler, R., *Proc. Natl. Acad. Sci. USA,* **68,** 1598–1602 (1971).

255. Hendry, I. A. and Iversen, L. L., *Brain Res.,* **29,** 159–162 (1971).

256. Otten, V. and Thoenen, H., *J. Neurochem.,* **29,** 69–75 (1977).

257. Otten, V. and Thoenen, H., *Mol. Pharmacol.,* **12,** 353–361 (1975).

258. Samuels, H. H. and Tomkins, G. H., *J. Mol. Biol.,* **52,** 57–74 (1970).

259. Munck, A. and Brinck-Johnsen, T., *J. Biol. Chem.,* **234,** 5556–5565 (1968).

260. Lamouroux, A., Faucon-Biguet, N., Samolyk, D., Privat, A., Salomon, J. C., Pujol, J. F., and Mallet, J., *Proc. Natl. Acad. Sci. USA,* **79,** 3881–3885 (1982).

261. Lewis, E. J., Tank, A. W., Weiner, N., and Chikaraisni, D. M., *J. Biol. Chem.,* **258,** 14632–14637 (1983).

262. Tank, A. W., Lewis, E. J., Chikaraishi, D. M., and Weiner, N., *J. Neurochem.,* **45,** 1030–1033 (1985).

263. Stachowiak, M., Sebbane, R., Stricker, E. M., Zigmond, M. J., and Kaplan, B. B., *Brain Res.,* **359,** 356–359 (1985).

264. Mallet, J., Biguet, N. F., Buda, M., Lamouroux, A., and Samolyk, D., *Cold Spring Harbor Symp. Quant. Bio., 48,* 305–308 (1983).

265. Black, I. B., Chikaraishi, D. M., and Lewis, E. J., *Brain Res., 339,* 151–153 (1985).

266. Tank, A. W., Curella, P., and Ham, L., *Mol. Pharmacol., 30,* 497–503 (1986).

267. Lewis, E. J., Harrington, C. A., and Chikaraishi, D. M., *Proc. Natl. Acad. Sci. USA, 84,* 3550–3554 (1987).

268. Short, J. M., Wynshaw-Boris, A., Short, H. P., and Hanson, R. W., *J. Biol. Chem., 261,* 9721–9726 (1986).

269. Karin, M., Haslinger, A., Holtgreve, H., Richards, R. I., Krauter, P., Westphal, H. M., and Beato, M., *Nature (London), 308,* 513–519 (1984).

270. Cambi, F., Fung, B., and Chikaraishi, D., *J. Neurochem., 53,* 1656–1659 (1989).

271. Curran, T. and Franza, B. R., Jr., *Cell, 55,* 395–397 (1988).

272. Vyas, S., Biguet, N. F., and Mallet, J., *EMBO J., 9,* 3707–3712 (1990).

273. Angel, P., Imagawa, M., Chiu, R., Stein, B., Imbra, R. J., Rahmsdorf, H. J., Jonat, C., Herrlich, P., and Karin, M., *Cell, 49,* 729–739 (1987).

274. Biguet, N. F., Vyas, S., and Mallet, J., *J. Physiol., 85,* 105–109 (1991).

275. Gizang-Ginsberg, E., and Ziff, E. B., *Genes Develop., 4,* 477–491 (1990).

276. D'Mello, S. R., Turzai, L. M., Gioio, A. E., and Kaplan, B. B., *J. Neurosci. Res., 23,* 31–40 (1989).

277. Fung, B. P., Yoon, S. O., and Chikaraishi, D. M., *J. Neurochem., 58,* 2044–2052 (1992).

278. Faucon-Biguet, N., Rittenhouse, A. R., Mallet, J., and Zigmond, R. E., *Neurosci. Lett., 104,* 189–194 (1989).

279. Schalling, M., Stieg, P. E., Linquist, C., Goldstein, M., and Hotkelt, T., *Proc. Natl. Acad. Sci. USA, 86,* 4302–4305 (1989).

280. Hefti, F., Gnahn, H., Schwab, M. E., and Thoenen, H., *J. Neurosci., 2,* 1554–1566 (1982).

281. Raynaud, B., Faucon-Biguet, N., Vidal, S., Mallet, J., and Weber, M. J., *Dev. Biol., 119,* 305–312 (1987).

282. Kilbourne, E. J. and Sabban, E. L., *Mol. Brain Res., 8,* 121–127 (1990).

283. Kilbourne, E. J., Nankova, B. B., Lewis, E. J., McMahon, A., Osaka, H., Sabban, D. B., and Sabban, E. L., *J. Biol. Chem., 267,* 7563–7569 (1992).

284. Kim, K.-S., Park, D. H., Wessel, T. C., Song, B., Wagner, J. A., and Joh, T. H., *Proc. Natl. Acad. Sci. USA, 90,* 3471–3475 (1993).

285. Gonzalez, G. A. and Montminy, M. R., *Cell, 59,* 675–680 (1989).

286. Sheng, M., McFadden, G., and Greenberg, M. E., *Neuron, 4,* 571–582 (1990).

287. Morgan, J. I. and Curran, T., *Nature (London), 322,* 552–555 (1986).

288. Sheng, M., Thompson, M. A., and Greenberg, M. E., CREB: *Science, 252,* 1427–1430 (1991).

289. Dash, P. K., Karl, K. A., Colicos, M. A., Prywes, R., and Kandel, E. R., *Proc. Natl. Acad. Sci. USA*, **88**, 5061–5065 (1991).

290. Nagatsu, T., *Cell. Mol. Neurobiol.*, **9**, 313–321 (1989).

291. Ichikawa, S., Ichinose, H., and Nagatsu, T., *Biochem. Biophys. Res. Commun.*, **173**, 1331–1336 (1990).

292. Craig, S. P., Buckle, V. J., Lamouroux, A., Mallet, J., and Craig, I., *Cytogenet Cell Genet.*, **42**, 29–32 (1986).

293. Grima, B., Lamouroux, A., Boni, C., Julien, J.-F., Javoy-Agid, F., and Mallet, J., *Nature (London)*, **326**, 707–711 (1987).

294. Kaneda, N., Kobayashi, K., Ichinose, H., Kishi, F., Nakazawa, A., Kurosawa, Y., Fujita, K., and Nagatsu, T., *Biochem. Biophys. Res. Commun.*, **146**, 971–975 (1987).

295. Kobayashi, K., Kaneda, N., Ichinose, H., Kishi, F., Nakazawa, A., Kurosawa, Y., Fujita, K., and Nagatsu, T., *J. Biochem.*, **103**, 907–912 (1988).

296. LeBourdelles, B., Boularand, S., Boni, C., Horellou, P., Dumas, S., Grima, B., and Mallet, J., *J. Neurochem.*, **50**, 988–991 (1988).

297. O'Malley, K. L., Anhalt, M. J., Martin, B. M., Kelsoe, J. R., Winfield, S. L., and Ginns, E. I., *Biochemistry*, **26**, 6910–6914 (1987).

298. Coker, G. T. I., Studelska, D., Harmon, S., Burke, W., and O'Malley, K. L., *Mol. Brain Res.*, **8**, 93–98 (1990).

299. Haycock, J. W., *J. Neurochem.*, **56**, 2139–2142 (1991).

300. Horellou, P., LeBourdellés, B., Clot-Humbert, J., Guibert, B., Leviel, V., and Mallet, J., *J. Neurochem.*, **51**, 652–655 (1988).

301. Kobayashi, K., Kiuchi, K., Ishii, A., Kaneda, N., Kurosawa, Y., Fujita, K., and Nagatsu, T., *FEBS Lett.*, **238**, 431–434 (1988).

302. Abate, C., Smith, J. A., and Joh, T. H., *Biochem. Biophys. Res. Commun.*, **151**, 1446–1453 (1988).

303. Abate, C., and Joh, T. H., *J. Mol. Neurosci.*, **2**, 203–215 (1991).

304. Kaufman, S., *Biochem. Soc. Trans.*, **13**, 433–436 (1985).

305. Ribeiro, P., Wang, Y., Citron, B. A., and Kaufman, S., *J. Mol. Neurosci.*, **4**, 125–139 (1993).

306. Liu, X. and Vrana, K. E., *Neurochem. Int.*, **1**, 27–31 (1991).

307. Hill, R. L., Hydrolysis of proteins, in *Advances in Protein Chemistry*, Vol. 20, Anfinsen, C. B. Anson, M. L. Edsall, J. T. and Richards, F. M., Eds., Academic, New York/London, 1965, pp. 37–107.

308. Kiuchi, K., Kiuchi, K., Titani, K., Fujita, K., Suzuki, K., and Nagatsu, T., *Biochemistry*, **30**, 10416–10419 (1991).

309. Ikeda, M., Fahien, L. A., and Udenfriend, S., *J. Biol. Chem.*, **241**, 4452–4456 (1966).

310. Iwaki, M., Phillips, R. S., and Kaufman, S., *J. Biol. Chem.*, **261**, 2051–2056 (1986).

311. Ledley, F. D., DiLella, A. G., Kwok, S. C. M., and Woo, S. L. C., *Biochemistry*, **24**, 3389–3394 (1985).

312. Fisher, D. B., Kirkwood, R., and Kaufman, S., *J. Biol. Chem.*, **247**, 5161–5167 (1972).

313. Kaufman, S., On the nature of an intermediate that is formed during the enzymatic conversion of phenylalanine to tyrosine, in *Iron and Copper Proteins, Advances in Experimental Medicine and Biology*, Vol. 74, Yasunobo, K. T. Mower, H. F. and Hayaishi, O. Eds., Plenum Press, New York/London. 1976, pp. 91–102.

314. Kaufman, S., Bridgers, W. F., Eisenberg, F., and Friedman, S., *Biochem. Biophys. Res. Commun.*, **9**, 497–502 (1962).

315. Daly, J., Levitt, M., Guroff, G., and Udenfriend, S., *Arch. Biochem. Biophys.*, **126**, 593–598 (1968).

316. Joh, T. H., Kapit, R., and Goldstein, M., *Biochim. Biophys. Acta*, **171**, 378–380 (1969).

317. Fitzpatrick, P. F., *Biochemistry*, **30**, 3658–3662 (1991).

318. Fitzpatrick, P. F., Chlumsky, L. J., Daubner, S. C., and O'Malley, K. L., *J. Biol. Chem.* **265**, 2042–2047 (1990).

319. Abita, J. P., Parniak, M., and Kaufman, S., *J. Biol. Chem.*, **259**, 14560–14566 (1984).

320. Fisher, D. B. and Kaufman, S., *J. Neurochem.*, **19**, 1359–1365 (1972).

321. Badaway, A. A., and Williams, D. L., *Biochem. J.*, **206**, 165–168 (1982).

322. Nicholls, P., and Schonbaum, G. R., Catalases, in *The Enzymes*, Vol. 8, Boyer, P. Lardy, H. and Myrbäck, K., Eds., Academic, New York, 1963, pp. 147–225.

323. Wallick, D. E., Bloom, L. M., Gaffney, B. J., and Benkovic, S. J., *Biochemistry*, **23**, 1295–1302 (1984).

324. Marota, J. J. A. and Shiman, R., *Biochemistry*, **23**, 1303–1311 (1984).

325. Taylor, R. J., Jr., Stubbs, C. S., Jr., and Ellenbogen, L., *Biochem. Parmacol.*, **18**, 587–594 (1969).

326. Bloom, L. M., Benkovic, S. J., and Gaffney, B. J., *Biochemistry*, **25**, 4204–4210 (1986).

327. *Critical Stability Constants*, Vol. 3, A. E. Martell and R. M. Smith, Ed., Plenum, New York/London, 1977, p. 200.

328. Dix, T. A. and Benkovic, S. J., *Biochemistry*, **24**, 5839–5846 (1985).

EXOPOLYSACCHARIDE ALGINATE SYNTHESIS IN *PSEUDOMONAS AERUGINOSA*: ENZYMOLOGY AND REGULATION OF GENE EXPRESSION

By SANDEEP SHANKAR, RICK W. YE, DAVID SCHLICTMAN, and A. M. CHAKRABARTY, *Department of Microbiology and Immunology (M/C 790), University of Illinois, College of Medicine, Chicago, Illinois*

CONTENTS

Advances in Enzymology and Related Areas of Molecular Biology, Volume 70, Edited by Alton Meister.
ISBN 0-471-04097-5   © 1995 John Wiley & Sons, Inc.

I. Introduction

Alginate is a linear random polymer of β-1,4-linked D-mannuronic acid and its C5 epimer L-guluronic acid. In some cases, the mannuronate residues are acetylated to various degrees by O-acetyl groups. Alginate is a polysaccharide of varied uses. It is widely used in the food, pharmaceutical, and chemical industry as a thickener, a gelling agent, or in the immobilization of various cells and enzymes (1, 2). For its commercial uses, alginate is manufactured mainly from marine brown algae, where as much as 40% of the dry cell material may be comprised of alginate. The major function of alginate in seaweeds appears to be mechanical strength and flexibility, depending on the guluronate content. High guluronate content generally provides high gel forming ability and mechanical rigidity to the tissues, while low guluronate alginate provides the flexibility for the seaweeds to float on water.

In addition to marine seaweeds, alginate is also produced by a number of microorganisms, namely, *Azotobacter vinelandii* and various pseudomonads such as *Pseudomonas aeruginosa, P. fluorescens,* and *P. mendocina* (3, 4). Alginate is primarily produced in *P. aeruginosa* during its infection in the cystic fibrosis (CF) lung (5). Alginate is also a major component of cysts, which are metabolically dormant cells, in *A. vinelandii* where it may account for as high as 70% of the intine and 40% of the exine carbohydrates. This polysaccharide coating is believed to protect the cells from desiccation and other stresses, and cysts have been reported to survive in dry soil for months or years (6). Alginate is also believed to protect brown algae from drying out in the littoral zone where they are exposed to air and high salt concentration during low tide (7). It is also interesting to note that a dehydrating agent such as ethanol triggers alginate synthesis in *P. aeruginosa* (8); since *P. aeruginosa* alginate is highly acetylated, which allows retention of moisture (9), it is likely that one function of alginate produced by *P. aeruginosa* during infection of the CF lung is to protect the infecting cells from the dehydrated environment of the lung, similar to many exopolysaccharide-producing bacteria that have a better survival advantage in a desiccated environment than the corresponding non-exopolysaccharide-producing strains (10, 11).

Protection from a dehydrated environment is not the only presumed function of alginate in *P. aeruginosa*. The alginate coating is believed to protect the infecting cells from antibiotic treatment and phagocytosis, as well as perhaps promote adherence to lung epithelial cells and formation of biofilms both on tissues and environmental surfaces (5, 12–14). Indeed, attachment to a solid surface in the form of a biofilm has been shown to promote exopolysaccharide formation, including alginate, in a number of microorganisms (15–17). Thus alginate formation is most likely a defense mechanism for *P. aeruginosa* in a stressed (dehydrated, high electrolyte) environment, such as the CF lung. Yet, there are a number of paradoxes in such an assumption. First, *P. aeruginosa* is not the only pseudomonad that is capable of producing alginate. Under appropriate conditions, a number of pseudomonads, not all human pathogens, such as *P. fluorescens, P. putida, P. mendocina,* and *P. syringae* (4, 18) can produce alginate, under circumstances that are not necessarily stressed. Second, alginate production in most pseudomonads is conditional, such that the biosynthetic genes are normally silent, but are specifically expressed under some environmental conditions. However, in the case of most pseudomonads including *P. aeruginosa,* when they undergo the genotypic transition to mucoidy (i.e., become alginate producer), they produce copious amounts of alginate that is presumably excessive for protection purposes. Third, a major signal in triggering transition to mucoidy, not only in *P. aeruginosa* but in many exopolysaccharide-producing bacteria, is starvation for nutrients. As will be emphasized in this chapter, nutrient limitations, particularly for phosphate or nitrogen, trigger an alarm system in *P. aeruginosa* that allows enhanced gene expression for some key energy-yielding reactions. Somehow, the trigger for these reactions is also geared to polysaccharide production. Thus the colonization and chronic infections in the CF lung by mucoid *P. aeruginosa* is more a reflection of the state of membrane perturbation and energy drain than the need for the bacteria to resist antibiotics or phagocytosis. In this chapter, we would like to present our current understanding on the nature of enzymes involved in alginate biosynthesis and modification, and discuss how the energy status of the cells may trigger specific expression of genes encoding the operation of the tricarboxylic acid (TCA) cycle and energy transduction. However, how enhanced operation of the TCA cycle and energy metabolism is linked to a genotypic transition to mucoidy is not clear at present.

II. Biosynthetic Enzymes

A. PHOSPHOMANNOSE ISOMERASE–GUANOSINE DIPHOSPHO-D-MANNOSE PYROPHOSPHORYLASE (PMI–GMP)

The pathway for alginate biosynthesis is shown in Figure 1(A). Fructose 6-phosphate (F6P) is ultimately converted to GDP-mannuronate (GDP–MA) residues that are polymerized, epimerized, and acetylated to form alginate. The first enzyme in this pathway is PMI–GMP.

The PMI–GMP is a bifunctional enzyme that catalyzes the conversion of F6P to mannose 6-phosphate (M6P) and mannose 1-phosphate (M1P) to GDP–mannose (GDP–M), the first and the third step of the alginate biosynthetic pathway (19). The PMI–GMP is encoded by the *algA* gene located in the biosynthetic gene cluster under the control of *algD* promoter [Fig. 1(B)]. This gene was originally isolated by complementing an *Escherichia coli* Pmi⁻ mutant strain. Normally, PMI and GMP activities in *P. aeruginosa* are at the border of detection limits in crude extracts, making the purification of this enzyme extremely difficult. However, when the *algA* gene is overexpressed under the *tac* promoter, both PMI and GMP activities are greatly elevated, facilitating the purification of this enzyme. The

Figure 1.   (A) Shows the pathway of alginate biosynthesis in *P. aeruginosa*. Abbreviations are as follows: F6P = fructose 6-phosphate; M6P = mannose 6-phosphate; GDP–Man = GDP–mannose; GDP–ManA = GDP–mannuronic acid; ACP = acyl-carrier protein. The symbols such as A, C, D, E, F, G, L, 8, 44, and 60 identify the genes in Fig. 1(B) that encode the enzymes. The enzyme nomenclatures (PMI, PMM, etc.) are shown near their corresponding gene products. Note that three of the enzymes, the lyase, acetylase, and epimerase, are located in the periplasm. The Alg60, whose function is unknown, has an acyl carrier protein domain at its carboxyl terminal and is believed to facilitate the transport of acetyl coenzyme A (CoA) to the acetylase in the periplasm. (B) Depicts the map locations and organization of the alginate biosynthetic and regulatory genes on the *P. aeruginosa* chromosome. Regulatory genes are shaded. The direction of transcription is shown by an arrow above each gene. The relative positions and orientations of AlgR1-binding sites upstream of the *algD* and *algC* genes are identified by open triangles.

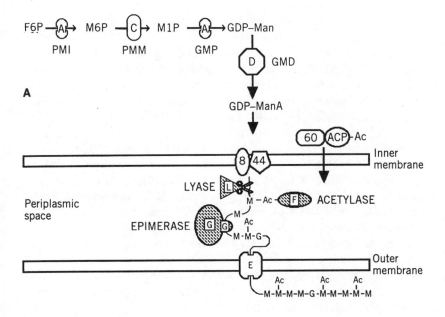

A

B

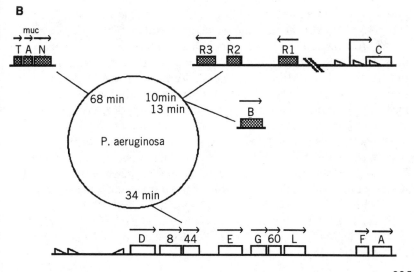

225

purification of PMI–GMP was initiated with hydrophobic interaction chromatography, followed by ion exchange, and gel filtration chromatography. The purified enzyme yielded a single polypeptide band on an SDS–PAGE gel with a molecular weight of 56 kDa under denaturing conditions. The N-terminal sequence determined from the protein matched the sequence predicted from the gene sequence. During the purification steps, the ratio of specific activities of PMI and GMP remained constant and the folds of purification were about the same for both enzymatic activities. Consequently, the presence of both activities was deemed not due to contaminations. Further evidence supporting the bifunctionality of the *algA* gene product was obtained by cloning the *algA* gene from the *P. aeruginosa* Alg⁻ strain 8853. The DNA sequencing showed that a single G to A base change at nucleotide 961 replaces Val^{321} with Met, resulting in a dramatic decrease in both PMI and GMP enzymatic activities. This mutation can be complemented with a *Bam*HI-*Sst*I fragment containing only the *algA* open reading frame. In addition, overexpression of this open reading frame led to simultaneous appearance of PMI and GMP activities in cell extracts.

The enzymatic activities of PMI and GMP were cold sensitive and addition of 15% glycerol to all purification buffers stabilized the enzyme and improved enzymatic activity. However, the enzyme tended to precipitate during concentration of column fractions. This problem was alleviated by adding NaCl to a final concentration of 200 mM whenever the enzyme was concentrated. The PMI activity of the enzyme showed a high sensitivity to reducing agents such as dithiothreitol (DTT). Addition of 0.1 mM of DTT resulted in 50% inactivation. The PMI activity was restored when the DTT was removed by dialysis. The GMP activity was not affected by up to 5 mM of DTT, suggesting that some of the sulfhydryl groups must remain oxidized for maximal PMI activity, but not for GMP. Despite the obvious sensitivity of PMI to DTT, enzyme stability during purification is maintained only when the reducing agent was present. Reducing agents were routinely removed before the enzymatic activity measurements. Characterization of the purified enzyme revealed that divalent metals are absolutely necessary for both PMI and GMP activity. For PMI activity, the order of activation is $Co^{2+} > Ni^{2+} > Mn^{2+} > Mg^{2+} > Ca^{2+} > Zn^{2+}$. However, the GMP reaction utilized either Mg^{2+} or Mn^{2+}. The K_m and V_{max} values for the reverse reaction of PMI with M6P as substrate were 3.03 mM and 830 nmol/min/mg of enzyme, respectively. The kinetic parameters for the forward reaction of PMI have not been determined due to the lack of an assay system. The forward reaction of GMP has a K_m value of 20 μM for mannose 1-phosphate (M1P) in the presence of 1 μM of guanosine 5′-triphosphate (GTP) and a K_m value of 30 μM for GTP in the presence of 1 μM of M1P. The K_m value for the reverse reaction with GDP–mannose (GDP–M) as substrate was determined to be 14 μM. It appears that the reaction equilibrium is more favorable for the forward reaction of PMI and for the reverse reaction of GMP.

A comparison of PMI–GMP with other isomerases and pyrophosphorylases revealed a weak homology. Two regions of PMI–GMP were, however, found to be quite similar to ADP–glucose pyrophosphorylase. First, the Lys[175] region (FVEKP) of PMI–GMP is identical to the substrate-binding site of bacterial ADP–glucose pyrophosphorylase and is highly conserved among the AlgA and other nucleotide sugar pyrophosphorylases, such as RfbM, CpsB, and XanB proteins (20). This result suggests that the amino acid sequence FVEKP may be a substrate-binding motif for this class of pyrophosphorylases [Fig. 2(A)]. Second, the Lys[20] region of PMI–GMP is similar to the allosteric site of ADP–glucose pyrophosphorylase. Mutation of Lys[175] of PMI–GMP to arginine, glutamine, or glutamate produced an enzyme whose K_m for M1P was 470–3200-fold greater than that measured for the wild-type enzyme (20). In addition, these mutant enzymes had a lower V_{max} for GMP activity. These results indicate that Lys[175] is primarily involved in the binding of the substrate M1P, although it is likely that other residues are required for the specificity of binding. Mutation of Arg[19] to glutamine, histidine, or leucine resulted in a twofold lower V_{max} for the GMP enzyme activity and a four- to sevenfold increase in the K_m for GTP compared with the wild-type enzyme. Thus, it appears that Arg[19] functions in the binding of GTP.

Mutations made through site-directed mutagenesis failed to selectively affect PMI activity. Chymotryptic digestion of PMI–GMP produced a major proteolytic fragment of 52 kDa that had low PMI activity but almost normal levels of GMP activity. Furthermore, the chymotryptic fragment could be selectively protected from further digestion by coincubation with the GMP substrates. The amino acid

A.

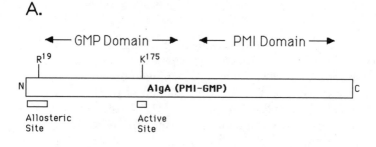

B.

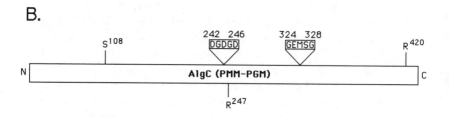

C.

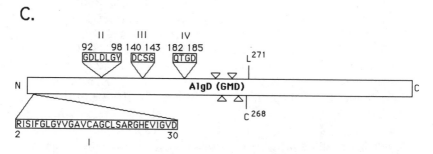

Figure 2. A functional peptide map of (A) PMI–GMP, (B) PMM/PGM, and (C) GMD. For the PMM/GM protein, the Ser[108] residue represents the putative active site that undergoes phosphorylation and dephosphorylation during the mutase reaction. The DGDGD loop is the putative divalent metal-binding site. The Arg[247] may be involved in the interaction with the substrate phosphate group. For the GMD protein, element I represents the putative NAD-binding site. Elements II, III, and IV represent the putative-binding site for the guanosine moiety of substrate GDP-D-mannose. Amino acid residues C[268] and L[271] may be part of active site involved in the catalysis of the substrate. The initial proteolytic cleavage sites are indicated by arrows.

sequence obtained with this proteolytic fragment indicated that chymotrypsin cleaved about 1 kDa from the carboxyl terminus of PMI–GMP. This result showed that the carboxyl terminus is critical for PMI activity but not for GMP activity. Taken together, these results support the hypothesis that the bifunctional PMI–GMP protein is composed of two independent enzymatic domains [Fig. 2(A)]. However, changes at Ser^{12} and Val^{321} reduced both PMI and GMP activities, suggesting that some parts of the protein are important for both functions.

B. PHOSPHOMANNOMUTASE/PHOSPHOGLUCOMUTASE (PMM/PGM)

The phosphomannomutase (PMM) enzyme is involved in the interconversion of M6P and M1P in the second step of the alginate biosynthetic pathway [Fig. 1(A)]. During the early stages of investigations on the enzymes involved in alginate biosynthesis, it was found that overexpression of *algA* gene led to an increase of PMM activity. Using this approach, a mutant strain, 8858, was found to lack PMM activity even in the presence of overexpressed *algA* gene. By complementation, the *algC* gene encoding PMM was subsequently cloned and sequenced (21). The amino acid sequence of PMM demonstrates homology to that of rabbit phosphoglucomutase (PGM) in some active domains, although the overall homology is marginal. The *algC* gene was then overexpressed under control of the *tac* promoter in pMMB66HE. The overexpressed enzyme was then purified in three steps: ion-exchange, hydrophobic interaction, and gel filtration chromatography (22). The purified enzyme migrated as a single polypeptide with a molecular mass of 52 kDa under denaturing conditions. The native protein was estimated to have a molecular mass of 47 kDa by gel filtration. Thus, the enzyme is a monomer. During initial attempts at purification of the protein, the specific activity decreased with subsequent purification steps. Ethylenediaminetetracetic acid (EDTA) treatments restored the enzymatic activity, both in crude extracts, as well as in collected fractions from different purification steps.

Phosphomannomutase exhibits a specific requirement for Mg^{2+} for activity. The EDTA treated enzyme has no activity if Mg^{2+} is absent. Manganese dichloride ($MnCl_2$) was only about 10% as effective at activating the enzyme. No activity was observed with other divalent metals or monovalent Li^+. The presence of other divalent metals besides Mg^{2+}, however, can effectively inhibit the enzymatic activity. Calcium dichloride ($CaCl_2$) has been shown to inhibit the enzyme activity competitively with a K_i value of 0.01 mM, while the K_m value for $MgCl_2$ is 0.4 mM. Monovalent Li^+ has only a slight inhibitory effect. The inhibitory effects of these divalent metals likely explain the enhanced effect of EDTA on PMM activity.

The purified enzyme can efficiently catalyze the interconversion of M1P and M6P, as well as that of glucose 1-phosphate (G1P) and glucose 6-phosphate (G6P). The apparent K_m values for M1P and G1P are 17 and 22 μM, respectively. On the basis of K_{cat}/K_m ratio, the catalytic efficiency for G1P was about twofold higher than that for M1P. Thus, the enzyme is in fact a phosphoglucomutase/phosphomannomutase (PGM/PMM). The dual activity is also supported by the observation that mutations in *algC* gene lead to a loss in both enzymatic activities in *P. aeruginosa*. Upon introduction of the wild-type *algC* gene, both activities were restored. The PGM/PMM can also catalyze the conversion of ribose 1-phosphate and 2-deoxyglucose 6-phosphate to their corresponding isomers, although activities are much lower (22).

Functional comparisons between PGM/PMM from *P. aeruginosa* and PGM from rabbit muscle (23) reveal significant similarities and differences. Both enzymes require glucose 1,6-diphosphate as cofactor and Mg^{2+} for maximum activity. They can use phosphoglucose, phosphomannose, and phosphoribose as substrates. The reaction equilibria for both enzymes are most favorable for conversion of C1 phosphosugars to C6 phosphosugars. However, the enzymes differ in their activities towards G1P and M1P. The PGM from rabbit muscle has a much higher K_{cat}/K_m value for G1P (2550 min^{-1} μM^{-1}) than that for M1P (5 min^{-1} μM^{-1}), suggesting that it is very specific for G1P. On the other hand, the K_{cat}/K_m values of PGM/PMM from *P. aeruginosa* are 136 min^{-1} μM^{-1} for G1P and 79 min^{-1} μM^{-1} for M1P. There is only a twofold difference with the *P. aeruginosa* enzyme instead of the 500-fold difference found with rabbit muscle PGM. The other difference between these two enzymes is that rabbit muscle PGM can be activated by Ni^{2+} and Co^{2+}, but PGM/PMM from *P. aeruginosa* is not. Reasons for this difference in divalent metal requirements are unknown.

Based on the functional and amino acid sequence comparisons

between the PGM/PMM from *P. aeruginosa* and the PGM from rabbit muscle (24), a functional peptide map of PGM/PMM can be constructed [Fig. 2(B)]. The Ser^{108} in domain I is involved in the phosphorylation and dephosphorylation during the reaction. The -(242)DGDGD(246)- loop in domain II is presumably involved in the binding of Mg^{2+} or other divalent metal ions. The Arg^{247} residue at the right side of the metal-binding loop is believed to interact with the phosphate group of Ser^{108}. The loop of -(324)GEMSG(328)- in domain III is most likely responsible for interacting with the sugar ring of the substrates. The Arg^{420} in domain IV may be responsible for interacting with the phosphate group of the substrates. There are other unique structural features of PGM/PMM that are not present in rabbit PGM. These include -(14)-FRAYDIR(20)- near the N-terminus and the phospho-gripper -(180)GNGVAG-(185)- . The functions of these loops are unknown.

The dual functions of PGM/PMM from *P. aeruginosa* reflect its dual roles in the biosynthesis of both alginate and lipopolysaccharide. The *algC* mutant of strain 8858 is incapable of alginate production due to the loss of PMM activity, while the *algC* mutant in strain PAO1 lacks O-side chain and core lipopolysaccharides (LPS) due to the loss of PGM activity. Besides PGM/PMM from *P. aeruginosa*, both PGM and PMM activities have been demonstrated in XanA from *Xanthomonas campestris* and PGM from *Neisseria gonorrhoreae*. The amino acid sequences of these three proteins are very similar. Similarities were also found in other genes encoding PGM and/or PMM from *E. coli*, *Salmonella typimurium*, and *Vibrio cholerae* (22). These enzymes form a subclass of phosphohexomutases. They all seem to play an important role in the biosynthesis of exopolysacharides and may have both PGM and PMM activities. The structural signature for this group of enzymes is the putative sugar ring binding loop: -GEMSG(A)- in domain III of the protein. Another unique feature is the presence of an ariginine residue (Arg^{420} in case of AlgC) in domain IV of the protein. Interestingly, this is the residue that was mutated in *algC* mutant strain 8858. The PGM enzymes that are specific for G1P, represented by PGM from rabbit muscle, *Agrobacterium tumefaciens* (GeneBank accession L24117) and *Acetobacter xylinum* (GeneBank accession L24077) form another subclass of phosphohexomutases. This subclass of enzymes has the GEESF(A)G loop as the interacting site for sugar ring of the substrates as determined by crystal structure. The third potential sub-class of phosphohexomutases are those enzymes that are specific for M1P. Such an enzyme has been reported in rabbit brain (25) but has not been found in bacteria so far.

C. GUANOSINE DIPHOSPHO-D-MANNOSE DEHYDROGENASE (GMD)

Guanosine diphospho-D-mannose dehydrogenase is a four-elec-tron-transfer dehydrogenase that catalyzes the conversion of GDP-mannose to GDP-mannuronic acid. This enzyme is encoded by the *algD* gene and is thought to commit the cell to alginate synthesis. However, like the enzyme PMI–GMP, the enzymatic activity of GMD is very low even in the heavily mucoid strains. This problem was solved again by overexpressing the gene under the *tac* promoter. The overexpressed protein was purified by heating the extract to 57.5°C for 10 min at a low pH, immediately followed by precipitation with 45% acetone and gel filtration (26). The protein has a subunit molecular weight of 48 kDa. The native molecular weight is 290 kDa and thus the enzyme is a hexamer. The enzyme is very unstable in pure form, but is stable for about 1 month at 4°C when suspended in a concentrated form after acetone precipitation. Although the en-zyme has an activity optimum at 50°C, activity was abolished within 2–3 min of raising the temperature above the optimum. The K_m and V_{max} values for substrate GDP-D-mannose are 14.9 μM and 581 nmol/min/mg, respectively. The enzyme appears to be specific for GDP-D-mannose since it did not utilize D-mannose, UDP-D-mannose, UDP-D-glucose, TDP-D-glucose, CDP-D-glucose, GDP-D-glucose, UDP-D-xylose, ethanol, D-glyceraldehyde 3-phosphate, L-lactate, or L-histidinol as alternative substrates. The enzyme has an apparent K_m of 185 μM for NAD and does not appear to use NADP as a cofactor.

A number of nucleoside phosphates and sugars have been tested as inhibitors of GMD activity. Guanosine monophosphate is the most potent inhibitor identified and has a competitive K_i of 22.7 μM. Prein-cubation with either *p*-hydroxymercuribenzoate or iodoacetamide resulted in a loss of enzymatic activity. This inactivation can be reversed by treating the mixture with excess dithiothreitol, suggest-ing possible involvement of sulfhydryl groups (cysteine residues) in enzymatic activity.

A preliminary structure analysis of GMD has been made by limited proteolyses (27). Results from proteolysis studies indicate that the enzyme may fold into two domains, a 25–26-kDa N-terminal domain and a 16–17-kDa carboxyl terminal domain. Linking these two domains is an exposed region (Tyr^{278}–Arg^{295}) that is susceptible to proteolysis. The binding of GDP-D-mannose to a tryptic N-terminal 26-kDa polypeptide product of GMD was assayed by the equilibrium dialysis method using GDP-D-[U-^{14}C] mannose as radioactive substrate. The trypsin-treated sample containing mainly the amino terminal T26 fragment showed the same ability to bind GDP-D-mannose as the undigested enzyme. Heat denaturation of this T26 fragment almost totally eliminated binding. Removal of the carboxyl end of the protein did not appear to affect the binding affinity toward GDP-D-mannose. The binding of NAD was analyzed with the photoaffinity analog [^{32}P]arylazido-β-alanine-NAD. The NAD binding site was also found to be located in the amino-terminal domain of the protein. Despite the fact that the N-terminal domain of the protein harbors both substrate- and cofactor-binding sites, the catalytic activity is not found in this domain. Based on homology comparisons with two other four-electron-transfer dehydrogenases, Cys^{268} appears to be involved in catalysis. Replacement of this amino acid residue by serine via oligonucleotide-directed mutagenesis abolished over 95% of the original GMD activity, confirming the importance of this cysteine residue (27). Replacement of the Lys^{271} with glutamine also resulted in loss of enzymatic activity. It has been shown that lysine and cysteine residues in UDP–glucose dehydrogenase take part in the two sequential steps of reaction, oxidation of a primary alcohol to the corresponding aldehyde, and the oxidation of the aldehyde to the corresponding carboxylic acid. These residues in GMD may have similar roles. However, attempts to isolate the product of the one-step oxidation using the Cys^{268} negative mutant were not successful.

Based on the above analysis and the amino acid sequence comparisons, a functional peptide map of GMD has been constructed as shown in Figure 2(C). Region I is most likely the binding site for NAD since it retains all the characteristics of an essential Rossmann-fold structure. In addition, an affinity labeling experiment has suggested the N-terminal location of the NAD-binding site. Three regions (II, III, and IV) near the N-terminus may participate in the binding of substrate, GDP-D-mannose. The spacing allows folding to bring these three regions together in the three-dimensional (3-D) structure to form the guanine nucleotide-binding site. It is interesting to note that Cys^{268} and Lys^{271} are located in a region clustered with initial proteolytic sites that separate the protein into two major domains. It is possible that this region comprises part of the cleft where active sites, including Cys^{268} and Lys^{271}, are located.

Finally, essentially nothing is known about the mechanism of polymerization of the mannuronate residues and secretion of this polymer. A protein termed AlgE has been characterized that appears to be a periplasmic protein, but is tightly associated with the outer membrane (28, 29). The non-porin AlgE protein is believed to be involved in alginate secretion, but its mechanism of action is not known (29).

III. Alginate Modifying Enzymes

As mentioned previously, alginate is composed of D-mannuronic acid and its C5 epimer L-guluronic acid. The *P. aeruginosa* alginate is also highly acetylated at the mannuronate residues. An important question in the synthesis of alginate is where and when does the polymerization, epimerization, or acetylation take place? Very little is known about how the mannuronate residues from GDP–mannuronate are polymerized to a polymannuronate polymer. Two putative membrane-bound proteins Alg8 and Alg44 (30) are believed to be involved in the polymerization process. Recently Alg8 gene has been postulated to encode a processive β-glycosyl transferase enzyme that presumably transfers mannuronic acid moieties from GDP-mannuronic acid to a growing oligosaccharide chain (30a). Another protein Alg60, which is essential for alginate synthesis and has an acyl carrier protein (ACP) moiety as part of its carboxyl terminal (A. Boyd and S. Mukhopadhyay, unpublished observations), is also believed to be involved in alginate polymerization and presumably acts by facilitating the transfer of the acetyl group from Acetyl-CoA to a putative alginate acetylase, the product of the *algF* gene (31, 32). The detailed mechanism of acetylation of the mannuronate residues in alginate is not understood. Acetylation of the mannuronate residues protects them from epimerization, suggesting that acetylation may occur before epimerization. Acetylation may also greatly reduce the susceptibility of alginate to alginate lyases such as AlgL, thereby maintaining the integrity of the alginate even in the presence of the lyase enzyme. Acetylation of *P. aeruginosa* alginate is believed to enhance the water retention property of the alginate capsule (9), thereby protecting the cells from the dehydrated environment of the CF lung. Acetylation has also been implicated in protecting the infecting cells from neutrophils and lymphocyte actions (33), thereby also enhancing their survival from the body's defense mechanisms.

A major modification of alginate, other than acetylation, is the epimerization of the mannuronate residues at the C5 position by an epimerase AlgG. Unlike the well-known *A. vinelandii* extracellular mannuronan C5 epimerase(s), which catalyzes the Ca^{2+}-dependent conversion of mannuronic acid to guluronic acid at the polymer level with extensive guluronate (G) block structures (34), the *P. aeruginosa* mannuronan C5 epimerase does not require Ca^{2+} for its activity, does not produce G block structures, and is mostly periplasmic, rather than extracellular (35). This epimerase activity was markedly inhibited *in vitro* by the presence of O-acetyl groups on the mannuronate residues, indicating that the degree of acetylation controls the level of epimerization. An *algG* mutant produced polymannuronate that was still acetylated, suggesting that epimerization is not a prerequisite for acetylation (35). Exactly how acetylase and epimerase work together in *P. aeruginosa* to produce alginate, which is highly acetylated and lacks G block structures, is not known at this time. Recently, the mannuronan C5 epimerase gene from *A. vinelandii* has been sequenced, and expressed in *E. coli* (36), which indicates the presence in this gene of repeated nonameric motifs putatively involved in Ca^{2+} binding, which might also be responsible for a secretion mechanism that does not involve the cleavage of an N-terminal signal sequence. Indeed, preliminary work from the laboratory of Valla (36a) indicates that *A. vinelandii* may produce multiple forms of the epimerase, each with the specificity to form block G or alternating M (mannuronate), G, or random M and G structures. It is not unlikely that *P. aeruginosa* may have remnants of such epimerases, but only a single functional form capable of introducing random G residues next to the M residues. The absence of an epimerase activity in *P. aeruginosa* capable of introducing G blocks into the alginate structure, which allows better gel formation in the presence of Ca^{2+}, suggests that *P. aeruginosa* produces a heavily acetylated, poor gel forming alginate for survival or detoxification reasons in a niche that differs from that of *A. vinelandii* or the marine brown algae.

In addition to acetylation and epimerization, a major modification of the *P. aeruginosa* alginate is its structural integrity, as influenced by the presence of the alginate lyase AlgL. The AlgL is a degradative enzyme that catalyzes the eliminative cleavage of the 4-O-linked glycosidic bonds between the uronate residues resulting in the cleavage of alginate with the production of unsaturated sugar derivatives. The *algL* gene is in the middle of the alginate biosynthetic gene cluster. It has been sequenced and its gene product purified after its expression in *E. coli* (37, 38). An AlgL⁻ mutant is mucoid, suggesting that AlgL is not absolutely necessary for alginate synthesis, although it has been postulated to have an editing function to control the length of the alginate polymer or it may prime the polymerization reaction by providing short segments of polymannuronate (39).

One function of AlgL is believed to allow *P. aeruginosa* cells to slough off from a biofilm or growth film on an environmental or human tissue (such as CF lung) surface under stressed conditions so that such cells can establish themselves in new niches (40). Alginate has been shown to allow *P. aeruginosa* to anchor themselves on a growth surface, and alginate-negative nonmucoid cells show a higher level of cell detachment from a growth film than the mucoid cells. Regulated expression of alginate lyase in a growth film allows alginate degradation to low molecular weight alginate and a higher level of cell detachment, indicating its putative role in extended biofilm formation due to enhanced sloughing of the cells from a preformed biofilm (40).

IV.　Regulation of Alginate Gene Expression

An important feature of alginate production by *P. aeruginosa* is that many biosynthetic genes are normally silent, but are specifically activated in the unique environment of the CF lung. The CF lung is characterized by a sticky, dehydrated mucous as well as a salty intracellular environment. It is interesting to note that high levels of electrolytes (high osmolarity), a dehydrating agent such as ethanol, as well as other features, such as nutrient starvation and energy limitations, all contribute to the activation of key alginate biosynthetic promoters, such as *PalgC* or *PalgD* (39). The mode of regulation of alginate gene expression has been reviewed recently (39), and will not be discussed in detail, except for those areas that are emerging as new and informative.

A. ENVIRONMENTAL ACTIVATION OF KEY ALGINATE PROMOTERS

The relevance of the CF lung environment and the triggering of mucoidy in *P. aeruginosa* has long been noted and emphasized previously. The organization of alginate biosynthetic (*alg*) genes, as depicted in Figure 1(B), shows the clustering of most of the *alg* structural genes, namely, *algD, alg8, alg44, algE, algG, alg60, algL, algF,* and *algA,* at the 34-min region of the *P. aeruginosa* chromosome. The major promoter in this cluster is the *PalgD*, which is believed to regulate the whole gene cluster as an operon (41), even though many genes within this cluster may be regulated in part by weak, internal promoters (29, 31). One major biosynthetic gene, *algC,* is located outside this cluster and is regulated independently by its own promoter *PalgC*. Both *PalgC* and *PalgD* are activated by high medium osmolarity, a characteristic of the CF lung (42, 43), as well as during adherence on a solid surface as a biofilm, in contrast to the growth of the cells in planktonic forms in a liquid medium (15, 16). The *PalgD* is also significantly activated by starvation conditions, particularly during nitrogen or phosphate starvation (44), as well as by membrane-perturbing agents such as ethanol (8).

Near *algC* at the 10–13-min region of the chromosome is a cluster of regulatory genes *algR1, algR2, algR3, algB,* and so on [Fig. 1(B)]. The *algR1, algR2,* and *algR3* genes are also known as *algR, algQ,* and *algP* (44, 45). Both *algR1* and *algB* demonstrate sequence homology to the family of two-component response regulators (46–48) and are known to allow activation of the *PalgD*. The mucoid status of the cells allows a somewhat higher level of *algB* expression, which has been shown to be positively regulated by *algT* and a *P. aeruginosa* homolog of the integration host factor (49). The exact mode of action of the *algB* gene, or the nature of environmental signals it may respond to, are, however, presently unknown.

AlgR1 acts as a transcriptional regulator of not only the *algD* gene, but also of its own expression (50) and that of the neuraminidase gene *nanA* (51) and the *algC* gene (42). There are multiple AlgR1-binding sites (ABS) upstream of both the *algC* (52) and the *algD* genes (53,

54). While the three ABSs in the case of *algD* are either near or very far upstream of the promoter, the three ABSs responsible for *algC* promoter activation are both upstream and downstream of the *PalgC*, with one being located within the *algC* gene itself (52). Indeed, the ABSs have been shown to behave as eukaryotic enhancer elements in promoting transcription from the *algC* or *algD* promoters independent of their orientation, number of copies, or position within the promoter (52). There are also auxilliary proteins needed for *algC* or *algD* activation. The DNA gyrase, needed for introducing negative supercoiling in the promoter region, is important for *algC* and *algD* promoter action (42, 55), as are the DNA-bending proteins CRP and IHF for *algD* promoter activation (55–58). Similar to the histone-like IHF, another histone-like protein AlgR3, alternatively called AlgP, also regulates *algD* promoter activation, even though its mechanism of action is unclear at present (59, 60). The involvement of a number of DNA-bending proteins such as CRP, IHF, perhaps AlgR3, as well as the need for a supercoiled promoter, and the obligate requirement of far upstream binding sites for AlgR1, suggests that DNA looping is involved in the activation of *algD*, and perhaps also of *algC* promoter.

It should be noted that transcriptional activation is not the only mechanism of the regulation of *algC* and *algD* expression. In addition, both these genes have long leader sequences (244 bp sequence for *algC* and 367 bp sequence for *algD*) that are involved in posttranscriptional regulation of these genes (61). The presence of a 5′-untranslated leader region as well as the ribosomal RNA-binding site have been shown to be important for efficient translation of the *algC* messenger RNA. How the leader region modulates enhanced translation of the alginate genes remains unknown at present.

V. ROLE OF THE *algR2* GENE IN TCA CYCLE OPERATION, ENERGY METABOLISM, AND ALGINATE SYNTHESIS

One of the regulatory genes involved in alginate synthesis is *algR2*, also called *algQ*. AlgR2 is needed for alginate synthesis at 37°C. The sequence of the *algR2* gene (62) does not show any appreciable homology with any known gene in the gene bank. A gene *pfrA*, homologous to *algR2*, has been characterized in *P. putida* WCS358 involved in siderophore synthesis, but its exact role in siderophore synthesis has not been delineated (63). Even though the role of AlgR2 in alginate synthesis was suggested to be that of a protein kinase (64, 65), it was later shown (66) that AlgR2 itself is not a kinase, but positively regulates the expression of nucleoside diphosphate kinase (Ndk), which forms a complex with succinyl-CoA synthetase (Scs) in *P. aeruginosa* (67). Since both Ndk and Scs are known to undergo autophosphorylation, the level of autophosphorylated proteins in an *algR2* knock-out mutant was found to be greatly reduced. This finding was also confirmed by western blotting using antibody against purified Ndk. Commensurate with a decreased level of phosphorylated Scs in the *algR2* mutant, the level of Scs as measured enzymatically was greatly reduced, which led to a greatly reduced growth rate of the *algR2* mutant with most of the TCA cycle intermediates as sole sources of carbon. It was therefore concluded that AlgR2 is important in energy metabolism in *P. aeruginosa* by positively regulating the Ndk–Scs complex (66), since this complex presumably allows the adenosine triphosphate (ATP), generated through substrate level phosphorylation during operation of the TCA cycle, to be used for the generation of other nucleoside triphosphates within the cell for important cellular activities (66). During nutrient and energy starvation, which is a potent signal for triggering mucoidy in nonmucoid cells (68–70), the cells apparently try to generate as much energy as possible through efficient use of the TCA cycle by enhancing the expression of *algR2* gene, and somehow this modulation of *algR2* expression triggers a switch to mucoidy. Whether AlgR2 controls disparate cellular functions, such as alginate synthesis in *P. aeruginosa* or siderophore formation in *P. putida* (63), both secretable products, through direct modulation of energy metabolism via the Ndk–Scs complex, or indirectly through its putative action on key promoters regulating the formation of these secretable products is unknown at the present time. AlgR2 has recently been demonstrated to regulate either positively or negatively, the levels of a number of secretable virulence factors such as alginate, siderophore, rhamnolipid and extracellular protease in *P. aeruginosa* (66a). In addition to *algR2*, another gene *algH* has also been implicated in the regulation of Ndk levels in *P. aeruginosa* cells (66a). Interestingly, two disparate genes from *E. coli*, *rnk* and *sspA*, have been shown to complement an *algR2* knock-out mutant in *P. aeruginosa* to mucoidy and *rnk* has been shown to regulate Ndk levels in *E. coli*

(66b, c). How SspA, a starvation-inducible protein, allows complementation of the *P. aeruginosa algR2* knock-out mutant to mucoidy,
has not been delineated as yet; however, preliminary evidence indicates that SspA may regulate the levels of intracellular nucleoside
triphosphates (NTP) through regulation of an alternative kinase that
can substitute for Ndk in NTP formation (66c).

A. MULTIPLE REGULATORY PROTEINS/ENZYMES CONTROL NDK ACTIVITY AND ALGINATE SYNTHESIS

An interesting property of the *P. aeruginosa algR2* mutant is that
while it is nonmucoid at 37°C, it is mucoid (i.e., alginate positive)
at 30°C or slowly becomes mucoid at 37°C. This condition is not due
to reversion of the *algR2* mutation, since such mucoid cells typically
behave as nonmucoid at 37°C but becomes slowly mucoid at 30 or
37°C. To see if this slow reversion to mucoidy is due to an inefficient
but cross-talking kinase, as was postulated earlier (66), we isolated
an *algR2algH* double mutant by chemical mutagenesis of the *al-
gR2*::Cm knock-out mutant, which exhibited an Alg⁻ phenotype at
all temperatures, even on prolonged incubation. Since the *algR2*
mutant shows a greatly reduced level of phosphorylated Ndk (16
kDa) and Scs (α subunit, 33 kDa) on incubation with [γ-^{32}P]ATP
(66), it was of interest to see if *algH* might additionally control the
level of autophosphosphorylation of Ndk and Scs in *P. aeruginosa*.
The extent of Ndk autophosphorylation in wild-type, *algR2* mutant
and the *algR2algH* double mutant is shown in Figure 3. As can be
seen, the *algR2* mutant exhibits 10–15% of the wild-type level of
Ndk that can undergo phosphorylation, and this residual level of
Ndk is further reduced significantly in the *algR2algH* double mutant,
suggesting that *algH* controls slow alginate synthesis via its modulation of Ndk gene expression. Thus Ndk levels within *P. aeruginosa*
cells are controlled by at least two genes, *algR2* and *algH* (66a).

We previously mentioned that even though the level of Ndk was
severely reduced in the *algR2* strain, the Ndk activity, as measured
by the transfer of the terminal phosphate from [γ-^{32}P]ATP to other
nucleoside diphosphates was not appreciably reduced, implying that
there was one or more cross-talking kinases that could substitute for
Ndk (66). We additionally showed that such a kinase(s) was susceptible to inhibition by Tween 20 (66). Tween 20 was observed to have

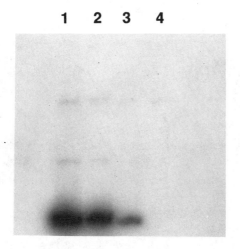

Figure 3. Autophosphorylation of nucleoside diphosphate kinase–succinyl CoA synthetase (Ndk–Scs) in crude extracts of wild-type *P. aeruginosa* strain 8830, the knock out *algR2*::Cm mutant, and the *algR2algH* double mutant. Some 10 μg samples of each of the sonicated cell free extracts of the three strains were incubated for 30 s with 100 nCi of [γ-³²P]ATP (3000 Ci/nmol) in a final volume of 20 μL. Reactions were stopped by the addition of 4 × SDS-PAGE buffer, electrophoresed on a 15% denaturing gel and analyzed by autoradiography. Lane 1: 10 μg of *P. aeruginosa* Ndk–Scs complex; Lane 2: wild-type 8830; Lane 3: 8830 *algR2*::Cm mutant; Lane 4: 8830 *algR2algH* double mutant. The faster moving more intense band is that of 16 kDa of Ndk while the less intense slower moving band is that of 33 kDa Scs (67).

no effect on Ndk and the cross-talking kinase(s) could not be blocked by anti-Ndk antibodies.

To see if the Tween 20-susceptible kinase(s) was the only cross-talking kinase(s), we examined the effect of both Tween 20 and anti-Ndk antibody, singly or in combination, on the phosphotransfer activity from [γ-³²P]ATP to other nucleoside diphosphates in cell-free extracts of the wild type, the *algR2* knockout mutant, and the *algR2algH* double mutant. The results are shown in Figure 4. As seen with SetA experiments, the transfer of terminal phosphate from [γ-³²P]ATP to UDP (with trace contaminations of CDP and GDP) as mediated by enzymes present in the crude extract of the wild-

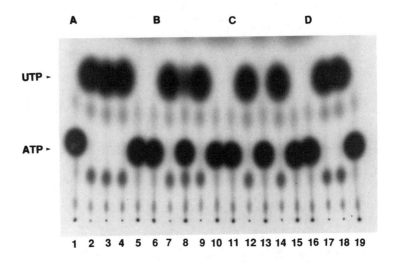

Figure 4. Analysis of nucleoside diphosphate to triphosphate conversion in crude extracts of wild-type *P. aeruginosa* strain 8830, 8830 *algR2*::Cm mutant and 8830 *algR2algH* double mutant in the absence or presence of Tween 20 and anti-Ndk antibody. Some 10 μg samples of sonicated extracts of the above three strains were incubated with 100 nCi of [γ-^{32}P]ATP (3000 Ci/nmol) and UDP at a final concentration of 2.5 μM with trace amounts of GDP and CDP. Reactions were conducted in the absence or in the presence of 0.1% Tween 20, 2 μL 1:1000-fold diluted anti-Ndk antibody, or both. Reactions were terminated after 30 s by the addition of 1 μL of 10% SDS. 1.0 μL of the reaction mixture was spotted onto a precoated polyethylene-imine plate and chromatographed in a 0.75 M KH$_2$PO$_4$ buffer, pH 4.5. Signals were visualized by autoradiography. The positions of ATP and UTP are marked. The GTP moves faster than ATP while CTP moves slower than ATP but faster than UTP, as shown. All reactions contain [γ-^{32}P]ATP. *Set A:* Lane 1: wild-type 8830 extract only; Lane 2: 8830 extract + UDP; Lane 3: 8830 extract + UDP + Tween 20; Lane 4: 8830 extract + UDP + anti-Ndk antibody; Lane 5: 8830 extract + UDP + Tween 20 + anti-Ndk antibody. *Set B:* Set B (lanes 6–10) is the same as Set A (lanes 1–5) except the extract was derived from 8830 *algR2*::Cm mutant. *Set C:* Set C (lanes 11–15) is the same as Set A (lanes 1–5) except the extract was derived from 8830 *algR2algH* double mutant. *Set D:* Set D experiments were conducted with purified Ndk, which was autophosphorylated in presence of [γ-^{32}P]ATP. Lane 16: Ndk; Lane 17: Ndk + UDP; Lane 18: Ndk + UDP + Tween 20; Lane 19: Ndk + UDP + anti-Ndk antibody.

type cell 8830 is not inhibited either by Tween 20 or by anti-Ndk antibodies (lanes 3 and 4) but is completely inhibited by a combination of the two (lane 5). This clearly suggests that there is a cross-talking kinase (or kinases) that is susceptible to inhibition by Tween 20. Inhibition of either Ndk or this cross-talking kinase cannot abolish phosphotransfer activity from [γ-^{32}P]ATP to other nucleoside diphosphates, but inhibition of both leads to complete inhibition of this activity. In the *algR2* mutant, where the Ndk activity is substantially reduced (Fig. 3, lane 3), inhibition of the cross-talking kinase by Tween 20 allows substantial, but not complete, inhibition of the phosphotransfer activity (Fig. 4, lane 8). Addition of anti-Ndk antibody, in addition to Tween 20, however, leads to a complete loss of the phosphotransfer activity (Fig. 4, lane 10). In contrast, the phosphotransfer activity in the *algR2algH* double mutant, which shows essentially no Ndk activity (Fig. 3, lane 4), is completely abolished either by Tween 20 alone (Fig. 4, lane 13) or by Tween 20 and anti-Ndk antibody (Fig. 4, lane 15), clearly indicating that either Ndk or the Tween 20 sensitive cross-talking kinase(s) is responsible for phosphotransfer activity from [γ-^{32}P]ATP to other nucleoside diphosphates in *P. aeruginosa*. Lanes 18 and 19 indicate that *P. aeruginosa* Ndk is not susceptible to Tween 20 inhibition, but of course is susceptible to inhibition by its own antibody.

An important question with regard to Ndk is its essentially to the cell. Since Ndk activity is primarily responsible for maintaining the nucleoside triphosphate/nucleoside diphosphate ratio within the cell, it has been shown to be essential for some organisms, while it is not in others, presumably because of cross-talking kinase activities (67). Since we have shown that there is essentially no Ndk activity in the *algR2algH* double mutant while the cells can still make nucleoside triphosphates from nucleoside diphosphates because of a Tween 20 sensitive cross-talking kinase(s), it was now possible for us to determine whether the overall nucleoside diphosphate kinase activity, as constituted by the 16-kDa Ndk and the cross-talking kinase(s), is essential for cell viability. The results are shown in Figure 5. In the absence of Tween 20, the wild type (WT), the *algR2* mutant (R2), and the *algR2algH* double mutant (R2H, which lacks the 16-kDa Ndk) grow normally, the latter two showing somewhat slower and lesser growth. In the presence of Tween 20, the growth of the R2 mutant is severely retarded while the growth of the R2H double

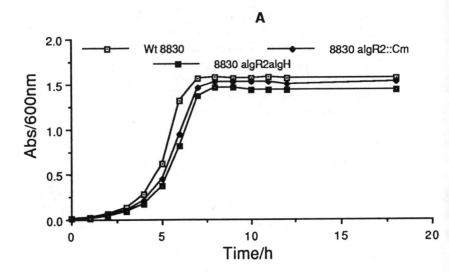

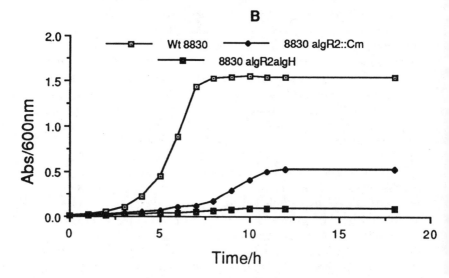

Figure 5.   Growth curves of *P. aeruginosa* wild-type strain 8830, 8830 *algR2*::Cm and the double mutant 8830 *algR2algH* in L broth in absence (set A, top) and in presence (set B, bottom) of 0.1% Tween 20.

mutant is almost completely abolished (66a, 66b). These results clearly suggest that while the 16-kDa Ndk is not absolutely essential to *P. aeruginosa* cells because of a cross-talking kinase, the inhibition of both the activities will abolish cellular growth, because of the importance of overall nucleoside diphosphate to nucleoside triphosphate conversion reactions. It should be noted that the *algR2algH* double mutant, in the absence of Tween 20, does not make any alginate at any temperature or growth phase, suggesting that Ndk, and not the cross-talking kinase, is important for alginate synthesis. How the *algR2algH* double mutant regulates Ndk levels or how the Ndk or the Ndk–Scs complex in turn controls alginate synthesis is presently unknown. Similarly, the nature or the number of the cross-talking kinase(s) has not been delineated as yet.

VI. Spontaneous Interconversion between Mucoidy and Nonmucoidy

It should be noted that transition of nonmucoid *P. aeruginosa* to mucoidy not only occurs predominantly in the CF lung, but can also be induced by hyperexpression of the *algT* gene (71), by mutations in the *muc* loci (72, 73), as well as during growth of the cells with ethanol (8) or with antibiotics such as carbenicillin (4) or kanamycin (74). This switch to mucoidy is not limited to *P. aeruginosa* either, and other pseudomonads, such as *P. putida* or *P. fluorescens,* can be induced to the alginate-positive state either through introduction of high copy numbers of *algT* genes or through growth on antibiotics (4, 71). Induction of mucoidy in *P. aeruginosa* is usually associated with loss of or reduction in the sizes of O side chains in the cell wall LPS molecules (75). The loss of such an O-antigen side chain may be due to activation of the *algC* or *algA* genes, which are also involved in LPS biosynthesis (22, 76), since cloned LPS genes are often known to affect LPS chain lengths (77). Since GDP–rhamnose, a major constituent of the LPS A band, is derived from GDP–mannose (78), and since in mucoid cells, activation of *PalgD* allows rapid conversion of GDP–mannose to GDP–mannuronic acid thereby draining the GDP–mannose pool, the transition to mucoidy may affect the overall synthesis of A and B band LPS in *P. aeruginosa*. In addition, transition to mucoidy has been reported to be associated with a reduced level of secretion of exoenzyme S, exotoxin A, phospholipase C, and pyochelin (70) as well as decreased level of extracellular proteases (79). It is thus possible that one or more genes involved in alginate synthesis may also regulate the production or secretion of exoenzymes or extracellular products (66a).

A. GLOBAL REGULATION OF ALGINATE PRODUCTION BY ALTERNATIVE SIGMA FACTOR, σ^E, ENCODED BY $algT$ ($algU$)

It has been demonstrated that typical nonmucoid strain PAO1 becomes mucoid upon the introduction of $algT$ ($algU$) gene *in trans* (71). Mutations in this gene lead to nonmucoid phenotypes in mucoid PAO derivatives. The amino acid sequence encoded by $algT$ ($algU$) reveals 66% identity with the alternate sigma factor σ^E from *E. coli*. The *P. aeruginosa* σ^E (AlgT) also shows similarities with *Myxococcus xanthus* CarQ, *P. syringae* HrpL, *Alcaligenes eutrophus* CnrH, *E. coli* FecI and SigX, and *Streptomyces coelicolor* SigE (80, 81). Phylogenetic analysis indicates that these proteins define a new subfamily of eubacterial RNA polymerase associated sigma factors that respond to extracytoplasmic stimuli and regulate extracytoplasmic functions. This new subclass of sigma factors is designated as the ECF subfamily of the σ^{70} family (80).

B. THE ROLE OF σ^E IN ALGINATE BIOSYNTHESIS

Sigma factors exert their functions by regulating specific promoters under their control. Although an *in vitro* transcriptional assay system has not yet been developed for $algD$ of *P. aeruginosa*, the promoter regions of genes involved in alginate production, including $algT$, have striking similarities with each other as well as those controlled by *E. coli* σ^E and *S. coelicolor* SigE, as shown in Figure 6. It has been postulated that $algT$, $algR1$, and $algD$, all are positively regulated by AlgT encoding a σ^E. Work from our laboratory (R. Ye, unpublished observations) has shown that the level of phosphomannomutase, the product of the $algC$ gene, is extremely low in an $algT$ mutant, suggesting that $algC$, which demonstrates putative σ^E binding motif (Fig. 6), is also under AlgT control.

The activation of the $algD$ promoter, which appears to control the whole biosynthetic gene cluster as an operon (41), has been shown to be under the control of $algT$. Two other genes located downstream of $algT$ [Fig. 1(B)] are involved in the negative regulation of σ^E:

-35 -10 $+1$

Ec	*rpoHP3*	CTTGCATT**GAACT**TGTGGA TAAAATCACGG**TCT**GATA**AAC**AGT
Sc	*dagAP2*	GCGTTCCG**GAACT**TTTGCACGCACGCGAGC**TCT**CGAATT**TT**GGC
Mx	*carQRS*	GAGCGCCGG**AAA**CACTTTCGCAGGTGGCCCG**T**AGAGGAG**T**CGGGT
Pa	*algT*	GGAGGGGA**GAACT**TTTGCAA GAAGCCCGAG**TCT**ATCTTGGCAAGACG
Pa	*algR*	GACTTGGGG**CACT**TTTCGGGCCTAAAGCGAG**TCT**CAGCGT**CG**
Pa	*algD*	AACGGCCGG**GAACT**TCCCTCGCAGAGAAAAC**ATC**CTATCACCGCGA
Pa	*algC*	ATCTTCAGG**AACT**CGGCGGGCAACGGACTGCCAAACCCCC**TGT**GCCT

Figure 6. Nucleotide sequence similarities of promoter regions under control of σ^E from different bacteria. Ec = *E. coli*; Sc = *Streptomyces coelicolor*, Mx = *Myxococcus xanthus*, Pa = *Pseudomonas aeruginosa*.

mucA and *mucB* (*algN*). Mutations in either *mucA* or *algN* result in mucoidy, suggesting MucA and AlgN together may function as anti-sigma factors or may repress *algT* expression. Expression of the *algD* promoter was increased in such a mutant background (81). When the gene cluster containing *algT, mucA, algN* was introduced into *E. coli* with the *algD-lacZ* fusion in the chromosome, an increase of *algD* promoter expression was observed (81). Mutation of either *mucA* or *algN* (*mucB*) resulted in further increase in *algD* expression. AlgT has also been shown to regulate the expression of *algB*, which is involved in the regulation of *algD* (49). These experiments further support the positive role of AlgT in the form of cascade regulation of alginate gene expression in *P. aeruginosa,* as proposed in Figure 7.

Strains of *P. aeruginosa* isolated from the initial stages of infection of CF patients are nonmucoid. With the progression of the disease, *P. aeruginosa* isolates become increasingly mucoid. The *mucA* gene plays a critical role in this conversion. Spontaneous mutations in *mucA* have been shown to be responsible for transition of nonmucoid *P. aeruginosa* cells to mucoidy (82). Complementation of these mutations results in suppression of alginate production and thus leads to a nonmucoid phenotype. Interestingly, it appears that many mucoid CF isolates have mutations in the *mucA* gene (Fig. 8), suggesting *mucA* is at least one of the hot spots of mutations responsible for conversion of nonmucoid *P. aeruginosa* to mucoidy in the CF lung (82). It will be worthwhile to further characterize the environmental signals and mechanisms that trigger these mutations in *mucA*.

The mucoid phenotype of *P. aeruginosa* isolated from the CF lung is often unstable, and nonmucoid variants can be observed, especially when cells are repeatedly passed on nonselected medium such as nutrient or L agar plates. An interesting question is if such transition to nonmucoidy is due to a reversion of mutations originally responsible for conversion to mucoidy (i.e., *mucA*). In contrast to conversion to mucoidy, suppression of mucoidy is most often due to mutations in *algT* (83). Several spontaneous mutations in *algT* gene leading to nonmucoid phenotype have been identified (Fig. 8). Complementation or gene replacement with the wild-type *algT* gene can often restore the mucoid phenotype.

Recently, we were able to clone the open reading frame of the *algT* gene and overexpress it in *E. coli* (84). Hyperproduced AlgT

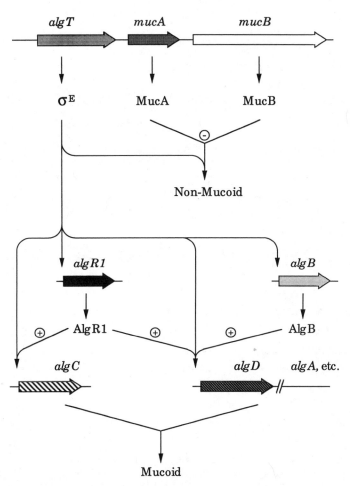

Figure 7. A model for the regulation of mucoidy–nonmucoidy interconversion. Either MucA or MucB is postulated to counteract the activity of AlgT, the putative RNA polymerase-associated σ^E, responsible for the transcription of *algR1, algB, algC,* and *algD* genes. Functional MucA or MucB is shown to produce a nonmucoid phenotype, while functional AlgT is shown to produce a mucoid phenotype through activation of the key alginate promoters.

A

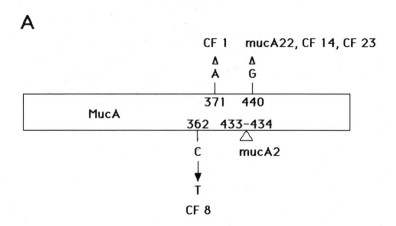

B

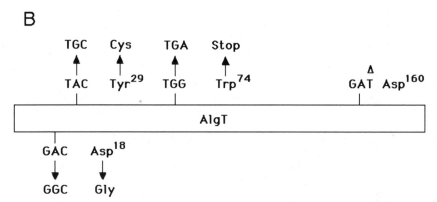

Figure 8.   Mutations in (A) *mucA* and (B) *algT* responsible for the mucoidy–nonmucoidy status of *P. aeruginosa*. Cystic fibrosis patient isolates are represented by CF.

was found mostly in inclusion bodies. AlgT was purified from the inclusion bodies by preparative SDS-PAGE and then renatured. The AlgT protein, in the presence of *E. coli* RNA polymerase core enzyme, allowed specific transcriptions from the σ^E-dependent *rpoH* promoter of *E. coli* and from the *algT* promoter itself. This provides biochemical evidence to support the notion that the *algT* gene encodes an alternate sigma factor (σ^E) which controls mucoidy in *P. aeruginosa*.

In summary, the *algT* gene encoding a σ^E controls the expressions of many genes involved in alginate production. Presence of putative anti-sigma factors, MucA and MucB (AlgN), suppresses the activity of AlgT although it is possible that MucA or AlgN may counter AlgT action by repressing *algT* gene expression. The mutations in the putative anti-sigma factor *mucA* may be responsible for conversion of many nonmucoid *P. aeruginosa* strains to mucoidy in the CF lung. On the other hand, mutations in *algT* under nonselective environments lead to suppression of mucoidy, resulting in a nonmucoid phenotype. Exactly how the CF lung environment may influence nonmucoid *P. aeruginosa* to undergo such mutations in *mucA,* or how growth in the absence of the CF lung allows spontaneous mutations in *algT* in mucoid CF isolates leading to nonmucoidy remains unknown at present.

Acknowledgments

Research in our laboratory has been supported by Public Health Service grants Al-16790-14 and Al-31546-02 from the National Institutes of Health, and in part by grants from the Cystic Fibrosis Foundation and Eli Lilly and Company. RWY is supported by a postdoctoral fellowship from the Cystic Fibrosis Foundation (F782).

References

1. Indergaard, M., *Appl. Phycol. Forum,* **8**(1), 2–4 (1991).
2. Skjak-Braek, G. and Martinsen, A., Applications of some algal polysaccharides in biotechnology, in *Seaweed Resources in Europe: Uses and Potential,* Guiry, M. D. and Blunden, G., Eds., Wiley, New York, 1991, pp. 219–257.
3. Pindar, D. F., and Bucke, C., *Biochem. J.,* **152**, 617–622 (1975).
4. Govan, J. R. W., Fyfe, J. A. M., and Jarman, T. R., *J. Gen. Microbiol.,* **125**, 217–220 (1981).
5. May, T. B., Shinabarger, D., Maharaj, R., Kato, J., Chu, L., DeVault, J. D., Roychoudhury, S., Zielinski, N. A., Berry, A., Rothmel, R. K., Misra, T. K., and Chakrabarty, A. M., *Clin. Microbiol. Rev.,* **4**, 191–206 (1991).
6. Sadoff, H. L., *Bacteriol. Rev.,* **39**, 516–539 (1975).

7. Skjak-Braek, G., *Biosynthesis and structure-function relationships in alginate.* University of Trondheim Publications, Trondheim, Norway, 1988, pp. 1–49.

8. DeVault, J. D., Kimbara, K., and Chakrabarty, A. M., *Mol. Microbiol., 4,* 737–745 (1990).

9. Skjak-Braek, G., Paoletti, S., and Gianferrara, T., *Carbohyd. Res., 185,* 119–129 (1989).

10. Roberson, E. B. and Firestone, M. K., *Appl. Environ. Microbiol., 58,* 1284–1291 (1992).

11. Ophir, T. and Gutnick, D. L., *Appl. Environ. Microbiol., 60,* 740–745 (1994).

12. Jensen, E. T., Kharazmi, A., Lam, K., Costerton, J. W., and Hoiby, N., *Infect. Immun., 58,* 2383–2385 (1990).

13. Bayer, A. S., Speert, D. P., Park, S., Tu, J., Witt, M., Nast, C. C., and Norman, D. C., *Infect. Immun., 59,* 302–308 (1991).

14. Mai, G. T., McCormack, J. G., Seow, W. K., Pier, G. B., Jackson, L. A., and Thong, Y. H., *Infect. Immun., 61,* 4338–4343 (1993).

15. Davies, D. G., Chakrabarty, A. M., and Geesey, G. G., *Appl. Environ. Microbiol., 59,* 1181–1186 (1993).

16. Hoyle, B. D., Williams, L. J., and Costerton, J. W., *Infect. Immun., 61,* 777–780 (1993).

17. Vandevivere, P. and Kirchman, D. L., *Appl. Environ. Microbiol., 59,* 3280–3286 (1993).

18. Fett, F. W., Osman, S. F., and Dunn, M. F., *Appl. Environ. Microbiol., 55,* 579–583 (1989).

19. Shinabarger, D., Berry, A., May, T. B., Rothmel, R., Fialho, A., and Chakrabarty, A. M., *J. Biol. Chem., 266,* 2080–2088 (1991).

20. May, T. B., Shinabarger, D., Boyd, A., and Chakrabarty, A. M., *J. Biol. Chem., 269,* 4872–4877 (1994).

21. Zielinski, N. A., Chakrabarty, A. M., and Berry, A., *J. Biol. Chem., 266,* 9754–9763 (1991).

22. Ye, R. W., Zielinski, N. A., and Chakrabarty, A. M., *J. Bacteriol, 176,* 4851–4857 (1994).

23. Ray, W. J. and Peck, E. J., Jr., Phosphomutase, in *The Enzymes,* Vol. 6, Boyer, P. D. Ed., Academic, New York, 1972, pp. 407–477.

24. Dai, J.-B, Liu, Y., Ray, W. J., Jr, and Konno, M., *J. Biol. Chem., 267,* 6322–6337 (1992).

25. Guha, S. K. and Rose, Z. B., *Arch. Biochem. Biophys, 243,* 168–173 (1985).

26. Roychoudhury, S., May, T. B., Gill, J. F., Singh, S. K., Feingold, D. S., and Chakrabarty, A. M., *J. Biol. Chem., 264,* 9380–9385 (1989).

27. Roychoudhury, S., Chakrabarty, K., Ho, Y.-K., and Chakrabarty, A. M., *J. Biol. Chem., 267,* 990–996 (1992).

28. Grabert, E., Wingender, J., and Winkler, U. K., *FEMS Microbiol. Lett., 68,* 83–88 (1990).

29. Chu, L., May, T. B., Chakrabarty, A. M., and Misra, T. K., *Gene*, **107**, 1–10 (1991).

30. Maharaj, R., May, T. B., Wang, S.-K., and Chakrabarty, A. M., *Gene*, **136**, 267–269 (1993).

30a. Saxena, I. M., Brown, R. M., Jr., Fevre, M., Geremia, R. A., and Henrissat, B., *J. Bacteriol*, **177**, 1419–1424 (1995).

31. Shinabarger, D., May, T. B., Boyd, A., Ghosh, M., and Chakrabarty, A. M., *Mol. Microbiol.*, **9**, 1027–1035 (1993).

32. Franklin, M. J. and Ohman, D. E., *J. Bacteriol.*, **175**, 5057–5065 (1993).

33. Mai, B. T., Seow, W. K., Pier, G. B., McCormack, J. B., and Thong, Y. H., *Infect. Immun.*, **61**, 559–564 (1993).

34. Skjak-Braek, G., Smidsrod, O., and Larsen, B., *Int. J. Biol. Macromol.*, **8**, 330–336 (1986).

35. Franklin, M. J., Chitnis, C. E., Gacesa, P., Sonesson, A., White, D. C., and Ohman, D. E., *J. Bacteriol.*, **176**, 1821–1830 (1994).

36. Ertesvag, H., Doseth, B., Larsen, B., Skjak-Braek, G., and Valla, S., *J. Bacteriol.*, **176**, 2846–2853 (1994).

36a. Ertesvag, H., Hoidal, H. K., Hals, I. K., Rian, A., Doseth, B., and Valla, S., *Mol. Microbiol.*, **16**, 000–000 (1995).

37. Boyd, A., Ghosh, M., May, T. B., Shinabarger, D., Keogh, R., and Chakrabarty, A. M., *Gene*, **131**, 1–8 (1993).

38. Schiller, N. L., Monday, S. R., Boyd, C. M., Keen, N. T., and Ohman, D. E., *J. Bacteriol.*, **175**, 4780–4789 (1993).

39. May, T. B. and Chakrabarty, A. M., *Trends Microbiol.*, **2**, 151–157 (1994).

40. Boyd, A. and Chakrabarty, A. M., *Appl. Environ. Microbiol.*, **60**, 2355–2359 (1994).

41. Chitnis, C. E. and Ohman, D. E., *Mol. Microbiol.*, **8**, 583–590 (1993).

42. Zielinski, N. A., Maharaj, R., Roychoudhury, S., Danganan, C. E., Hendrickson, W., and Chakrabarty, A. M., *J. Bacteriol.*, **174**, 7680–7688 (1992).

43. Berry, A., DeVault, J. D., and Chakrabarty, A. M., *J. Bacteriol.*, **171**, 2312–2317 (1989).

44. DeVault, J. D., Berry, A., Misra, T. K., Darzins, A., and Chakrabarty, A. M., **7**, 352–358 (1989).

45. Konyecsni, W. M. and Deretic, V., *J. Bacteriol.* **172**, 2511–2520 (1990).

46. Deretic, V., Dikshit, R., Konyecsni, W. M., Chakrabarty, A. M., and Misra, T. K., *J. Bacteriol.*, **171**, 1278–1283 (1989).

47. Wozniak, D. J. and Ohman, D. E., *J. Bacteriol.*, **173**, 1406–1413 (1991).

48. Goldberg, J. B. and Dahnke, T., *Mol. Microbiol.*, **6**, 59–66 (1992).

49. Wozniak, D. J. and Ohman, D. E., *J. Bacteriol*, **175**, 4145–4153 (1993).

50. Kimbara, K. and Chakrabarty, A. M., *Biochem. Biophys. Res. Commun.*, **164**, 601–608 (1989).

51. Cacalano, G., Kays, M., Saiman, L., and Prince, A., *J. Clin. Invest.*, **89**, 1866–1874 (1992).

52. Fujiwara, S., Zielinski, N. A., and Chakrabarty, A. M., *J. Bacteriol.*, **175**, 5452–5459 (1993).

53. Kato, J. and Chakrabarty, A. M., *Proc. Natl. Acad. Sci. USA*, **88**, 1760–1764 (1991).

54. Mohr, C. D., Leveau, J. H. J., Krieg, D. P., Hibler, N. S., and Deretic, V., *J. Bacteriol.*, **174**, 6624–6633 (1992).

55. DeVault, J. D., Hendrickson, W., Kato, J., and Chakrabarty, A. M., Mol. Microbiol. **5**, 2503–2509 (1993).

56. Mohr, C. D. and Deretic, V., *Biochem. Biophys. Res. Commun.*, **189**, 837–844 (1992).

57. Toussaint, B., Delic-Attree, I., and Vignais, P. M., *Biochem. Biophys. Res. Commun.*, **196**, 416–421 (1993).

58. Wozniak, D. J., *J. Bacteriol.*, **176**, 5068–5076 (1994).

59. Kato, J., Misra, T. K., and Chakrabarty, A. M., *Proc. Natl. Acad. Sci. USA* **87**, 2887–2891 (1990).

60. Deretic, V., Mohr, C. D., and Martin, D. W., *Mol. Microbiol.*, **5**, 1577–1583 (1991).

61. Fujiwara, S., and Chakrabarty, A. M., *Gene,* **146**, 1–5 (1994).

62. Kato, J., Chu, L., DeVault, J. D., Kimbara, K, Chakrabarty, A. M., and Misra, T. K., *Gene,* **64**, 31–38 (1989).

63. Venturi, V., Ottevanger, C., Leong, J., and Weisbeek, P. J., *Mol. Microbiol.*, **10**, 63–73 (1993).

64. Roychoudhury, S., Sakai, K., Schlictman, D., and Chakrabarty, A. M., *Gene,* **112**, 45–51 (1992).

65. Roychoudhury, S., Sakai, K., and Chakrabarty, A. M., *Proc. Natl. Acad. Sci. USA,* **89**, 2659–2663 (1992).

66. Schlictman, D., Kavanaugh-Black, A., Shankar, S., and Chakrabarty, A. M., *J. Bacteriol.*, **176**, 6023–6029 (1994).

66a. Schlictman, D., Kubo, M., Shankar, S., and Chakrabarty, A. M., *J. Bacteriol.*, **177**, 2469–2474 (1995).

66b. Schlictman, D., Shankar, S., and Chakrabarty, A. M., *Mol. Microbiol.*, **16**, 309–320 (1995).

66c. Shankar, S., Schlictman, D., and Chakrabarty, A. M., Mol. Microbiol., **16**, (1995), in press.

67. Kavanaugh-Black, A., Connolly, D. M., Chugani, S. A., and Chakrabarty, A. M., *Proc. Natl. Acad. Sci. USA,* **91**, 5883–5887 (1994).

68. Speert, D. P., Farmer, S. W., Campbell, M. E., Musser, J. M., Selander, R. K., and Kuo, S., *J. Clin. Microbiol.*, **28**, 188–194 (1990).

69. Terry, J. M., Pina, S. E., and Mattingly, S. J., *Infect. Immun.*, **59**, 471–477 (1991).

70. Woods, D. E., Sokol, P. A., Bryan, L. E., Storey, D. G., Mattingly, S. J., Vogel, H. J., and Ceri, H., *J. Infect. Dis.,* **163,** 143–149 (1991).

71. Goldberg, J. B., Gorman, W. L., Flynn, J. L., and Ohman, D. E., *J. Bacteriol.,* **175,** 1303–1308 (1993).

72. Fyfe, J. A. M. and Govan, J. R. W., *J. Gen. Microbiol.,* **119,** 443–450 (1980).

73. Ohman, D. E. and Chakrabarty, A. M., *Infect. Immun.,* **33,** 142–148 (1981).

74. Deretic, V., Tomasek, P., Darzins, A., and Chakrabarty, A. M., *J. Bacteriol.,* **165,** 510–516 (1986).

75. Hancock, R. E. W., Mutharia, L. M., Chan, L., Darvean, R. P., Speert, D. P., and Pier, G. B., *Infect. Immun.,* **42,** 170–177 (1983).

76. Goldberg, J., Hatano, K., and Pier, G. B., *J. Bacteriol.,* **175,** 1605–1611 (1993).

77. Lightfoot, J. and Lam, J. S., *J. Bacteriol.,* **173,** 5624–5630 (1991).

78. Lightfoot, J. and Lam, J. S., *Mol. Microbiol.,* **8,** 771–782 (1993).

79. Ohman, D. E. and Chakrabarty, A. M., *Infect. Immun.,* **37,** 662–669 (1982).

80. Lonetto, M. A., Brown, K. L. Rudd, K. E., and Buttner, M. J., *Proc. Natl. Acad. Sci. USA,* **91,** 7573–7577 (1994).

81. Schurr, M. J., Martin, D. W., Mudd, M. H., and Deretic V., *J. Bacteriol.,* **176,** 3375–3382 (1994).

82. Martin, D. W., Schurr, M. J., Mudd, M. H., Govan, J. R. W., Holloway, B. W., and Deretic V., *Proc. Natl. Acad. Sci. USA,* **90,** 8377–8381 (1993).

83. DeVries, C. A. and Ohman, D. E., *J. Bacteriol.,* **176,** 6677–6687 (1994).

84. Hershberger, C. D., Ye, R. W., Parsek, M. R., Xie, Z. D., and Chakrabarty, A. M., *Proc. Natl. Acad. Sci. USA,* **92,** (1995), in press.

AUTHOR INDEX

Numbers in parentheses are reference numbers and indicate that the author's work is referred to although his name is not mentioned in the text. Numbers in *italics* indicate pages on which the complete reference appears.

257

SUBJECT INDEX

279

CUMULATIVE AUTHOR INDEX

291

308 CUMULATIVE AUTHOR INDEX

CUMULATIVE SUBJECT INDEX

VOL. PAGE

VOL. PAGE